Residential Design Using
Autodesk Revit Building 8

Daniel John Stine

ISBN: 1-58503-234-4

SDC

PUBLICATIONS

Schroff Development Corporation

www.schroff.com

www.schroff-europe.com

Foreword

To The Student:

This book has been written with the assumption that the reader has no prior experience using Autodesk Revit. It would be beneficial if the student has some basic computer knowledge and understanding of how to manage files using Microsoft Windows; however, it is not absolutely necessary. The intent of this book is to provide the student with a well-rounded knowledge of tools and techniques for use in both school and industry.

It is strongly recommended that this book is completed in lesson order. All exercises utilize drawings created in previous lessons.

To the Instructor:

This book was designed for the architectural student using Autodesk Revit Building 8. Throughout the book the student develops a single family residence. The drawings start with the floor plans and develop all the way to photo-realistic renderings like the one on the cover of this book.

Throughout the book many Revit tools and techniques are covered while creating the house model. Also, in a way that is applicable to the current exercise, industry standards and conventions are covered. Access to the internet is required for some exercises.

An Instructor's resource guide is available with this book. It contains:
- Answers to the questions at the end of each chapter
- Outline of tools & topics to be covered in each lesson's lecture
- Suggestions for additional student work (for each lesson)

About the Author:

Dan Stine has thirteen years experience in the architectural field. He currently works at LHB (a 140 person multidiscipline firm; www.LHBcorp.com) in Duluth Minnesota as an Architectural Technical Designer. Dan has worked in four firm's total. While at these firms, he has participated in collaborative projects with several other firms on various projects (including Cesar Pelli, Weber Music Hall – University of Minnesota - Duluth). All of these firms have their own CAD standards and customization. This has given Dan a fairly well rounded knowledge of optimizing and implementing CAD standards, customization and document organization. Dan is a Construction Documents Technician (CDT), a member the Construction Specification Institute (CSI) and the Autodesk Developer Network (ADN). He teaches CAD classes at Lake Superior College, for the Architectural Technology program. Mr. Stine has also written books titled "*Introduction to Autodesk Revit Building 8*" (which is a tutorial book, similar to this one, based on a commercial project) and "*Residential Design Using AutoCAD 2005*", both of which are also Published and distributed by SDC Publications.

You can contact Dan with comments or suggestions at **dan.stine@charter.net**
Please do not email with Revit questions unless they relate to a problem with this book.

Thanks:

I could not have done this with out the support from my family; Cheri, Kayla & Carter. They had to bear with me a few nights while daddy had to work on the book, again. I love you guys!

Many thanks go out to Stephen Schroff and Schroff Development Corporation for making this book possible!

Table of Contents

1. **GETTING STARTED WITH AUTODESK REVIT BUILDING 8**   Page
    1-1   What is Autodesk Revit Building 8?   1-1
    1-2   Overview of the Revit user interface   1-3
    1-3   Open, Save & Close an existing Project   1-8
    1-4   Creating a new Project   1-13
    1-5   Using Pan & Zoom to view your drawings   1-15
          Self-Exam & Review Questions

2. **Lake Cabin: FLOOR PLAN (The Basics)**
    2-1   Walls   2-1
    2-2   Doors   2-5
    2-3   Windows   2-7
    2-4   Roof   2-9
    2-5   Annotation & Dimensions   2-11
    2-6   Printing   2-15
          Self-Exam & Review Questions

3. **Overview of Linework & Modify Tools**
    3-1   Lines and Shapes   3-1
    3-2   Snaps   3-16
    3-3   Edit Tools   3-22
    3-4   Annotations   3-35
          Self-Exam & Review Questions

4. **Drawing 2D Architectural Objects**
    4-1   Sketching Rectilinear Objects   4-1
    4-2   Sketching Objects with Curves   4-11
          Self-Exam & Review Questions

5. **Residence: FLOOR PLAN (First Floor)**
    5-1   Project setup   5-1
    5-2   Exterior walls   5-7
    5-3   Interior walls   5-23
    5-4   Doors Openings and Windows   5-35
    5-5   Adding a fireplace   5-54
          Self-Exam & Review Questions

6. **Residence: FLOOR PLANS (Second Floor & Basement Plans)**
    6-1   View setup and enclosing the shell   6-1
    6-2   Adding the interior walls   6-5
    6-3   Adding Doors, Openings & Windows   6-9
    6-4   Basement Floor Plan   6-13
    6-5   Stairs   6-20
    6-6   Dimensions   6-33
          Self-Exam & Review Questions

7. **Residence: ROOF**
    7-1   Roof Design options   7-1
    7-2   Gable roof   7-11
    7-3   Low roof elements   7-21
    7-4   Skylights   7-32
          Self-Exam & Review Questions

8. **Residence: FLOOR SYSTEMS & REFLECTED CEILING PLANS**
    8-1   Floor systems   8-1
    8-2   Ceiling systems (Susp. ACT & Gypsum Board)   8-15
    8-3   Placing Light Fixtures   8-30
    8-4   Annotations   8-35
          Self-Exam & Review Questions

9.   Residence: ELEVATIONS   **Page**
9-1   Creating & viewing parametric exterior elevations   9-1
9-2   Modifying the Project model: exterior elevations   9-9
9-3   Creating & viewing parametric interior elevations   9-16
9-4   Modifying the Project model: interior elevations   9-21
9-5   Design Options   9-24
       Self-Exam & Review Questions

10.   Residence: SECTIONS
10-1   Specify section cutting plane in plan view   10-1
10-2   Modifying the Project model in section view   10-7
10-3   Wall Sections   10-14
10-4   Annotations & Detail components   10-17
       Self-Exam & Review Questions

11.   Residence: FLOOR PLAN FEATURES
11-1   Bathroom layouts   11-1
11-2   Kitchen Layout   11-7
11-3   Furniture   11-20
11-4   Adding Guardrails   11-25
        Self-Exam & Review Questions

12.   Residence: SCHEDULES
12-1   Room & Door tags   12-1
12-2   Generate a Door Schedule   12-6
12-3   Generate a Room Finish Schedule   12-10
       Self-Exam & Review Questions

13.   Residence: PHOTO-REALISTIC RENDERING
13-1   Creating an exterior rendering   13-1
13-2   Rendering an isometric in section   13-13
13-3   Creating an interior rendering   13-20
13-4   Adding people to the rendering   13-27
       Self-Exam & Review Questions

14.   Residence: CONSTRUCTION DOCUMENTS SET
14-1   Setting up a sheet   14-1
14-2   Sheet Index   14-10
14-3   Printing a set of drawings   14-17
       Self-Exam & Review Questions

**Partial view of
project created
in this book**

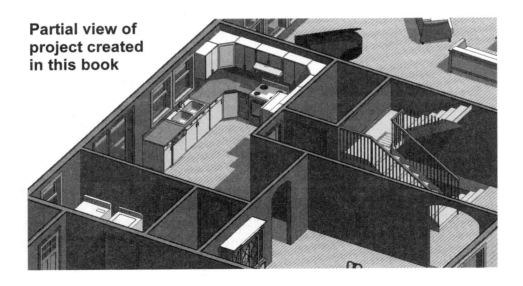

Lesson 1
Getting Started with Autodesk Revit Building 8::

This chapter will introduce you to Autodesk Revit Building 8. You will study the User Interface and learn how to open and exit a project and adjust the view of the drawing on the screen. It is recommended that the student spend an ample amount of time learning this material, as it will greatly enhance your ability to progress smoothly through subsequent chapters.

Exercise 1-1:
What is Autodesk Revit Building 8?

What is Autodesk Revit Building 8 used for?

Autodesk Revit Building 8 is the world's first fully parametric architectural design software. This revolutionary software, for the first time, truly takes architectural computer aided design beyond simply being a high tech pencil. Autodesk Revit Building 8 is a product of Autodesk, makers of AutoCAD, Architectural Desktop and 3D Studio Max. The Autodesk company web site claims more than 4 million users (of its various products) in 160 countries. Autodesk's 3,600 employees create products available in 20 languages.

What is a parametric building modeler?

Revit is a relatively new program designed from the ground up using state-of-the-art technology. The term parametric describes a process by which an object is modified in one view and automatically updated in all other views and schedules. For example, if you move a door in an interior elevation view, the floor plan will automatically update. Or, if you delete a door, it will be deleted from all other views and schedules. You can even delete a door from the door schedule and the drawings will instantly be revised to reflect the change.

A major goal of Revit is to eliminate much of the repetitive and mundane tasks traditionally associated with CAD programs to allow more time for design and visualization. For example, all sheet numbers, elevation targets and reference bubbles are updated automatically when changed anywhere in the Project. It is impossible to have a miss-referenced detail tag.

The best way to understand how a parametric model works is to describe the Revit project file. A single Revit file contains your entire building project. Even though you mostly draw in 2D views, you are actually drawing in 3D. In fact, the entire building project is a 3D model. From this 3D model you can generate 2D elevations, 2D sections and perspective views. Therefore, when you delete a door in an elevation view you are actually deleting the door from the 3D model from which all 2D views are generated (and automatically updated). The Freedom Tower in NYC is being designed by SOM using Revit and is a single file!

Why use Revit?

Many people ask the question, why use Revit versus other programs? The answer can certainly vary depending on the situation and particular needs of an individual/organization.

Generally speaking, this is why most companies use Revit:
- Many designers and drafters are using Revit to streamline repetitive drafting tasks and focus more on designing and detailing a project.
- Revit is a very progressive program and offers many features for designing buildings. Revit is constantly being developed and Autodesk provides incremental upgrades/patches on a regular basis; this version was released less than a year after the previous version.
- Revit was designed specifically for architecture and includes features like:
  - Accurender's Photo-realistic renderer
  - Phasing *(which makes Revit a 4D tool; time being the 4th dimension)*
  - Pantone digital color
  - Vectoral Shadows *(real-time shadows)*
  - Design options

With the recent acquisition of Revit by Autodesk and the statement on their web site that this is the path they will develop long-term, Revit is quickly becoming the industry standard in architectural design and documentation. With a solid company like Autodesk behind Revit, a prospective user can be fairly certain that the investment in time and resources will ultimately pay off.

The future is limitless.

We can expect some amazing advancements in the program as it develops. For example, with the building being a 3D model, we will see building code analyzers with plug-in modules for state and local codes, similar to tax programs like TurboTax. We might also see structural, mechanical and electrical engineers designing in the same model as the architects. **Autodesk Revit Structural** has just been released! This would eliminate conflicts found in many drawings.

Exercise 1-2:
Overview of the Revit User Interface

Revit is a powerful and sophisticated program. Because of its powerful feature set it has a measurable learning curve, though its intuitive design makes it easier to learn than other CAD programs. However, like anything, when broken down into smaller pieces, we can easily learn to harness the power of Revit. That is the goal of this book.

This section will walk through the different sections of the User Interface (UI). As with any program, understanding the user interface is the key to using the programs features.

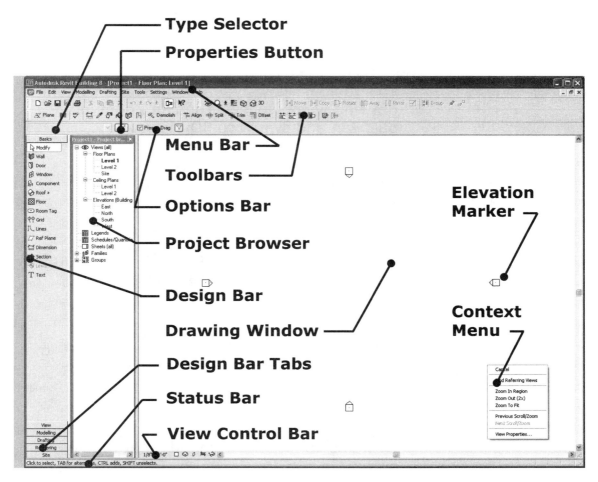

Figure 1-2.1 Revit User Interface

The Revit User Interface:

Menu bar:
Like all Windows programs, Revit has a series of pull-down menus across the top of the screen. Click on each of the menus to explore their contents. Many of these commands are graphically represented by Toolbars and the Design Bar.

Toolbar - Standard:
The Standard toolbar contains commands found in most Windows programs. Some examples are: Open, Save, Cut, Copy, Paste, Undo, Redo and Print.

new open save save* print   cut copy paste delete undo redo   browser what's this

Toolbar - View:
This toolbar allows you to adjust the current drawing window view. You can Zoom in and out, Pan and switch to 3D Views.

view zoom thinlines show mass 3D view

Toolbar - Edit:
Contains commands that modify objects in the drawings, i.e. Move, Mirror, Array, Group, Rotate.

resize   pin   create
                      similar

Toolbar - Tools:
Contains common commands that modify objects in the drawings, i.e. Trim, Split, Align, Tape Measure.

work plane, spelling, tape measure, match type, linework, paint, split face, edit cut profile, demo...

Toolbar - Worksets: This toolbar allows you to manage and work with Worksets. Worksets allow multiple users to work on the same project (i.e. same project file). This toolbar is not visible by default; you can turn it on by: *Window → Toolbar → Worksets* via the menu bar.

Toolbar –
   Design Options: This toolbar allows you to manage various Design Options in your project. One Design Option might be a hip vs. a gable roof. This toolbar is not visible by default; you can turn it on by: *Window → Toolbar → Design Options* via the menu bar.

Options Bar: This "toolbar" dynamically changes to show options that compliment the current operation.

Options Bar example with Wall tool active:

type selector   properties draw pick-lines   wall height
                                                   pick-faces

Project Browser: The Project Browser shows all the views, families, sheets, legends, schedules and groups available in the current project. A view is a floor plan, elevation or ceiling plan of the model.

Design Bar: This area is like an enhanced Menu Bar. Each tab (see Design Bar Tabs next) displays commands related to the tab title. This area also dynamically changes to show options related to the current operation.

Design Bar Tabs:

The Design Bar Tabs are groupings of commands in the Design Bar area. The current tab is the tab located at the bottom of the top group of tabs. Notice when you select a Design Bar Tab (e.g. Structural) that the tabs above it also slide to the top. Think of it like drawers of design tools.

TIP: Right-click on a Design Bar Tab to see a pop-up menu that allows you to turn on and off various tab groups.

Status Bar:

This area will display information about the current command or list information about a selected object.

TIP: Look here often while you are learning Revit.

Status Bar example with the Demolish tool active:

Drawing Window:

This is where you design your project and generate views and schedules.

Type Selector:

The Type Selector lists the available families for the current operation. For example, when inserting a door you can select different door styles and sizes; when inserting walls you can select various wall types to draw. This important feature is actually on the *Options Bar*.

Properties Button:

This button allows you to view the various properties of the selected components or active tool. Many of the properties are editable here as well. This feature is also apart of the *Options Bar*.

View Control Bar:

This is a feature introduced in version 7 which gives you convenient access to tools that control each views display settings (i.e. scale, shadows, detail Level and graphics style).

Context Menu: The Context Menu appears near the cursor
                                  whenever you right-click on the mouse. The
                                  options on that menu will vary depending on
                                  what tool is active or what is selected.

Context menu example with a wall selected:

Cancel
Flip Orientation
Select Previous   Ctrl+LeftArrow
Properties...
Make Worksets Editable
Make Elements Editable
Delete
Create Similar
Select All Instances
Edit Family
Find Referring Views
Zoom In Region
Zoom Out (2x)
Zoom To Fit
Previous Scroll/Zoom
Next Scroll/Zoom
View Properties...

Elevation Marker: This item is not really part of the Revit UI, but
                                  is visible in the drawing window by default (via
              the Default.rte template that Revit starts with),
                                  so it is worth mentioning at this point. The four
                                  elevation markers point at each side of your
                                  floor plan and ultimately indicate which
                                  drawing sheet you would find an elevation
                                  drawing of that side of the building. All you
                                  need to know right now is that you should
                                  draw your floor plan generally in the middle of
                                  the four elevation markers that you will see in
                                  each plan view.

This concludes our brief overview of the Revit user interface. Many of
these tools and operations will be covered in more detail throughout
the book.

Exercise 1-3:
Open, Save and Close an Existing Project

Open **Autodesk Revit Building 8:**
Start → All Programs → Autodesk → Autodesk Revit Building 8

NOTE: *The Windows XP Start menu is shown in Figure 1-3.1.*

Or double-click the Revit icon from your desktop.

This may vary slightly on your computer; see your instructor or system administrator if you need help.

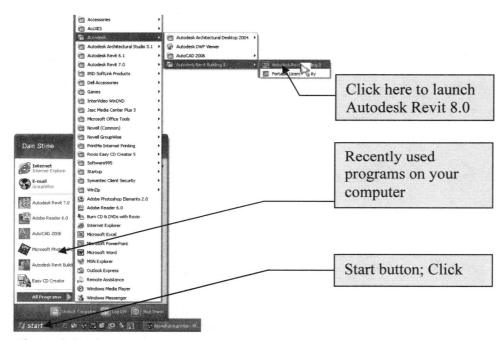

Figure 1-3.1 Starting Revit

Open an existing Revit project:

By default, Revit will open a new (empty) project. So, the first thing we will do is close this project.

1. Select **File → Close** from the menu bar.

Next you will open an existing Revit project file. You will select a sample file that was installed with the Revit program.

2.   Select **File → Open** from the menu bar.

3.   On the left side of the Open dialog box, scroll down and click on the **Training Files** icon (Figure 1-3.2):

   a.   **TIP**: *Alternatively you can browse via My Computer, to the training files:* **C:\Documents and Settings\All Users\Application Data\Autodesk\Revit Building 8\Training.** *If you do not see the Application Data folder, it may be because your folder is set to "hidden" via My Computer. In My Computer go to Tools → Folder Options → View tab and then check the option:* Show hidden files and folders.

   b.   **TIP**: *The training folder may not contain any files depending on how Revit was installed on your computer. You can accesses the files on the Revit CD or on Autodesk's web site (www.autodesk.com).*

Figure 1-3.2 Open dialog: click Training Files shortcut

4.   Double-click (with the left mouse button) the **Imperial** folder.

5.   Select the file named **i_Urban_House.rvt** and click **Open**.

The *i_Urban_House.rvt* file is now open and the last saved view is displayed in the Drawing Window (Figure 1-3.3).

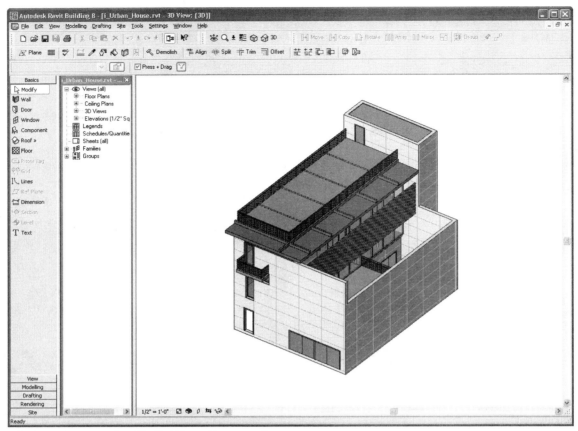

Figure 1-3.3 Training file "i_Urban_House.rvt"

The *Window* pull-down menu on the menu bar lists the projects and views currently open on your computer.

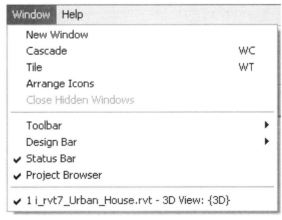

6. Click **Window** from the menu bar (Figure 1-3.4).

Notice that the *i_Urban_House.rvt* project file is listed. Next to the project name is the name of a view (e.g. floor plan, elevation) open on your computer.

Figure 1-3.4 Window menu

Additional views will be added to the list as you open them. Each view has the project name as a prefix. The current view (i.e. the view you are working in) has a check mark next to it. You can quickly toggle between opened views from this menu.

Open another existing Revit project:

Revit also lets you open more than one project at a time.

7. Click **File → Open** from the menu bar.

8. Per the instructions above, browse to **Training Files**.

9. In the **Imperial** folder, select the file named **i_Curtain_Walls.rvt** and click **Open** (Figure 1-3.5).

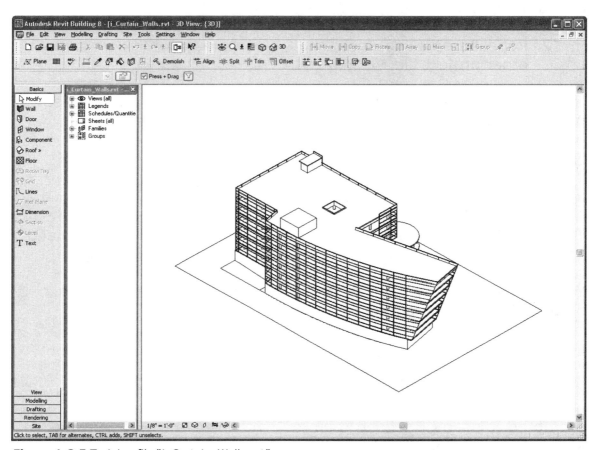

Figure 1-3.5 Training file "i_Curtain_Walls.rvt"

10. Click **Window** from the menu bar (Figure 1-3.6).

Notice that the *i_Curtain_Walls.rvt* project is now listed along with a view (*3D View: Southeast Isometric* in this example).

Try toggling between projects by clicking on *i_Urban_House.rvt – 3D View*.

Close a Revit project:

11. Select **File → Close** from the menu bar.

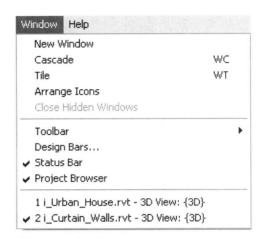

Figure 1-3.6 Window menu

This will close the current project/view. If more than one view is open for a project, only the current view will close. The project and the other opened views will not close (until you get to the last open view).

12. Repeat step 11 to close the other project file.

If you have not saved your view yet, you will be prompted to do so before Revit closes the view. **Do not save at this time**.

Saving a Revit project:

At this time we will not actually save a project.

To save a Project view, simply select **File → Save** from the file menu. You can also click the save icon from the Standard Toolbar.

You should get in the habit of saving often to avoid losing work due to a power outage or program crash.

Closing the Revit program:

Finally, from the *File* pull-down menu select *Exit*. This will close any open projects and views and shut Revit down. Again you will be prompted to save (if needed) before Revit closes the view. **Do not save at this time**.

You can also click the red X in the upper right corner of the Revit window. (*The icon is red in Windows XP only.*)

Exercise 1-4:
Creating a New Project

Open **Autodesk Revit**.

Creating a new project file:

The steps required to set up a new Revit building model project file are very simple. As mentioned earlier, simply opening the Revit program starts you in a new unnamed project file.

To manually create a new project (maybe you just finished working on a previous assignment and want to start the next one):

1. Select **File → New → Project...** from the file menu bar or select the *New* icon from the Standard toolbar.

If you select the *New* icon, a new project is quickly setup using the default template.

If you select *File → New → Project...* from the menu bar, you will get the **New Project** dialog box (Figure 1-4.1).

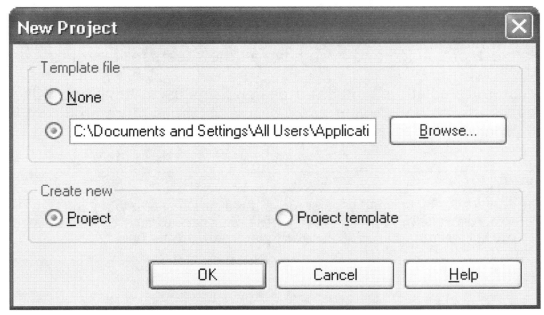

Figure 1-4.1 New Project dialog box

The *New Project* dialog box lets you specify the template file you want to use, or not use a template at all. You can also specify whether you want to create a new project or template file.

2. Leave the **default.rte** *Template file* selected (you need to click in the text box and "arrow key" to the right to see the template file name), and leave *Create new* set to **Project** (Figure 1-4.1).

3. Click **OK**; you now have a new "unnamed" project file.

To name an unnamed project file you simply save. The first time an unnamed project file is saved you will be prompted to specify the name and location for the project file.

4. Select **File → Save** from the menu bar.

5. Specify a **name** and **location** for your new project file.
   Your instructor may specify a location or folder for your files in this class.

What is a template file?

A template file allows you to start your project with certain settings preset the way you like or need them.

For example, you can have the units set to Imperial or Metric. You can have the door, window and wall families you use most loaded and eliminate others less often used. Also, you can have your company's title block preloaded and even have all the sheets for a project setup.

A custom template is a must for design firms using Revit and will prove useful to the student as he or she becomes more proficient with the program.

Be Aware:
It will be assumed from this point forward that the reader understands how to create, open and save project files. Please refer back to this section as needed. If you still need further assistance ask your instructor for help.

Exercise 1-5:
Using Zoom and Pan to View your Drawings

Learning to Pan and Zoom in and out of a drawing is essential to accurate and efficient drafting and visualization. We will review these commands now so you are ready to use them with the first design exercise.

Open **Autodesk Revit**.

We will select a sample file that was installed with the Revit program.

1. Select **File → Open** from the menu bar.

2. Browse to the **Training Files** area.
   (Reminder: Click the "Training Files" shortcut icon on the left.)

3. In the Imperial folder select the file named
   i_Drawing_exercise.rvt and click **Open** (Figure 1-5.1).

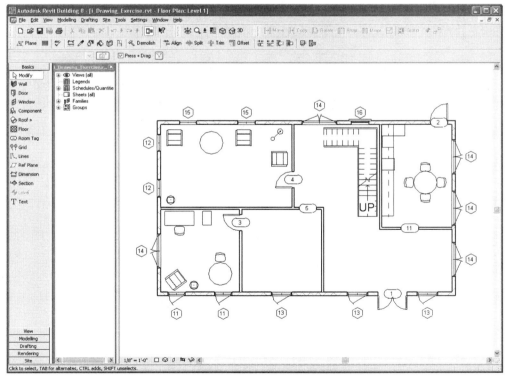

Figure 1-5.1 i_Drawing_exercise.rvt project

If the default view that is loaded is not **Floor Plan: Level 1**, double-click on **Level 1** under **Views\Floor Plans** in the *Project Browser*. Level 1 will be bold.

Using Zoom and Pan tools:

You can access the zoom tools from the **View** toolbar, or the *View* pull-down menu and the *scroll wheel* on your mouse.

View toolbar commands (from left to right):
- Dynamically modify view
- Zoom In *(includes drop-down arrow for additional zoom tools)*
- Thin Lines *(to be covered in a later lesson)*
- Show Mass
- Default 3D view

Zoom In

4.   Select the zoom icon by clicking directly on the magnifying glass (not the down-arrow).

5.   Drag a window over your plan view (Figure 1-5.2).

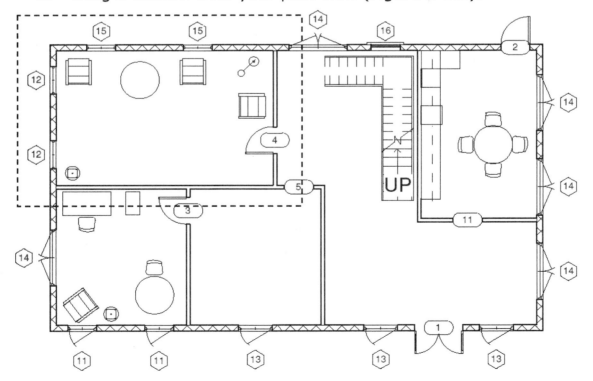

Figure 1-5.2 Zoom In window

You should now be zoomed in to the specified area (Figure 1-5.3).

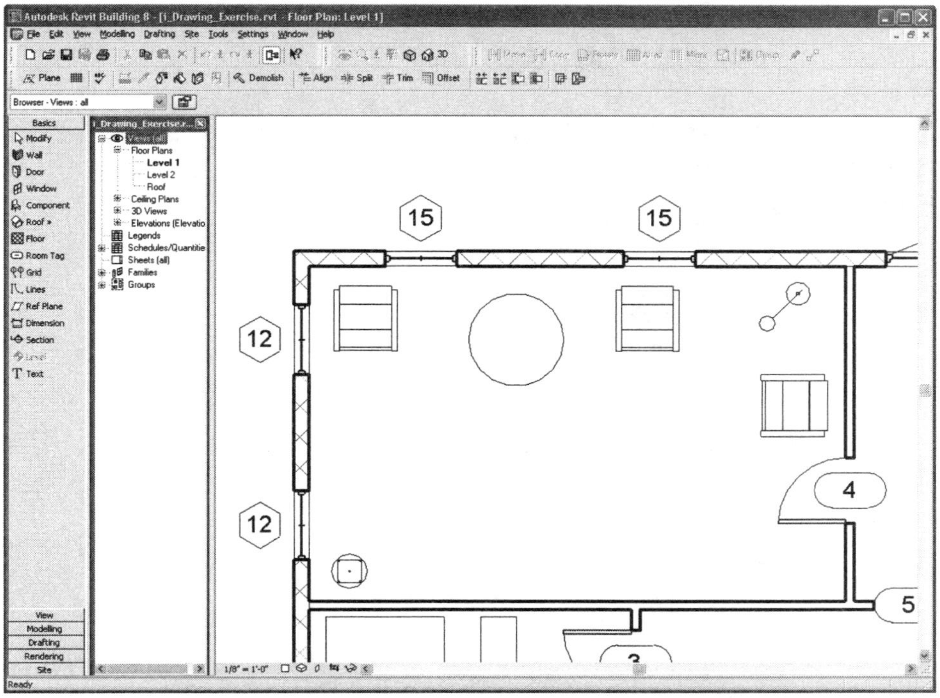

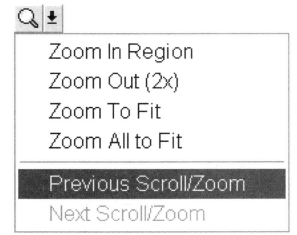

Figure 1-5.3 Zoom In results

Zoom Out

6. Click the down-arrow next to the *Zoom* icon (Figure 1-5.4). Select **Previous Scroll/ Zoom**.

You should now be back where you started.

Take a minute and try the other zoom tools to see how they work. When finished, click **Zoom All to Fit** before moving on.

Zoom In Region
Zoom Out (2x)
Zoom To Fit
Zoom All to Fit

Previous Scroll/Zoom
Next Scroll/Zoom

Figure 1-5.4 Zoom Icon drop-down

Dynamically Modify View

The *Dynamically Modify View* tool allows you to perform a real-time *Zoom* and *Pan* in your drawing view.

    7.   Click the **Dynamically Modify View** icon; you should see the *Dynamic View* dialog box at the lower left corner of your screen (Figure 1-5.5).

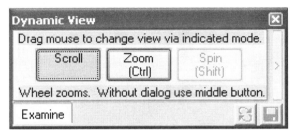

Figure 1-5.5 Dynamic View dialog box

    8.   You are automatically in *Scroll* mode. Drag your mouse (holding left button) around the screen in the drawing window to scroll (also called *panning*) the view of your drawing in the window.

> ***Be Aware:***
> You are not moving the drawing. You are just changing what part of the drawing you see in the drawing window.

Next you will try the real-time zoom. You can click the *Zoom* button in the *Dynamic View* dialog (Figure 1-5.5) or you can simply hold down the Control (Ctrl) key on the keyboard to toggle into Zoom mode. Using the Ctrl key is a quick way to switch back and forth between Scroll and Zoom.

    9.   Hold down the Ctrl key and drag your mouse up or down (vertically) in the drawing window.

FYI: *The Spin option is only available in 3D views. You will try this shortly.*

Default 3D View

Clicking on the *Default 3D View* icon loads a 3D View in another drawing window. This allows you to quickly switch to a 3D view.

10. Click on the **Default 3D View** icon.

11. Click on the **Window** pull-down menu and notice the 3D view and the Floor Plan view are both listed at the bottom.
    Remember: *You can toggle between views here.*

12. Click the **Esc** key to close the *Window* menu.

13. Click the **Dynamically Modify View** icon.
    Notice *the Spin button is now active (Figure 1-5.6).*

14. Click on the **Spin button** (Figure 1-5.6) *or hold down the Shift key*, then drag the mouse in any direction (horizontal or vertical) in the 3D view.

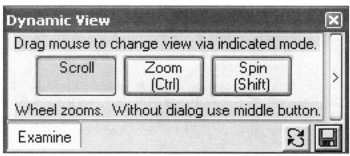

Figure 1-5.6 Dynamic View dialog box

15. **Close** the *i_drawing_exercise* project without saving.

Using the Scroll Wheel on the Mouse

The scroll wheel on the mouse is one of the best improvements to the computer in recent years. In Revit you can Pan and Zoom without even clicking a zoom icon. You simply **scroll the wheel to zoom** and **hold the wheel button down to pan**. This can be done while in another command (e.g. while drawing walls). Another nice feature is that the drawing zooms into the area near your cursor, rather than zooming only at the center of the drawing window like the Dynamic View tool does.

Self-Exam:

The following questions can be used as a way to check your knowledge of this lesson. The answers can be found at the bottom of this page.

1. The View Toolbar allows you to save your project file. (T/F)

2. You can zoom in and out using the wheel on a wheel mouse. (T/F)

3. Revit is a parametric architectural design program. (T/F)

4. A _____ file allows you to start your project with certain setting preset the way you like or need them.

5. In the Revit user interface, projects are viewed in the _____ window.

Review Questions:

The following questions may be assigned by your instructor as a way to assess your knowledge of this section. Your instructor has the answers to the review questions.

1. The Options Toolbar dynamically changes to show options that compliment the current operation. (T/F)

2. Revit is strictly a 2D drafting program. (T/F)

3. The Projects/Views listed at the bottom of the Window pull-down menu allow you to see which Projects/Views are currently open. (T/F)

4. When you use the scroll tool you are actually moving the drawing, not just changing what part of the drawing you can see on the screen. (T/F)

5. Revit was not originally created for architecture. (T/F)

6. The icon with the floppy disk picture ( ) allows you to _____ a project file.

7. Clicking on the _____ next to the 'Zoom In' icon will list additional zoom tools not currently shown in the view toolbar.

8. When using the Dynamically Modify View tool, the Spin button is inactive unless you are in a _____ view.

Lesson 2
Lake Cabin: FLOOR PLAN::

In this lesson you will get a down and dirty overview of the functionality of Revit. We will cover the very basics of creating the primary components of a floor plan: Walls, Doors, Windows, Roof, Annotation and Dimensioning. This lesson will show you the amazing "out-of-the-box" powerful, yet easy to use, features in Revit. It should get you very excited about learning this software program. Future lessons will cover these features in more detail while learning other editing tools and such along the way.

Exercise 2-1:
Walls

In this exercise we will draw the walls, starting with the exterior. Read the directions carefully, everything you need to do is clearly listed.

Tracing Paper Sketch of Lake Cabin Plan:

This sketch is just to give you an idea of what you will be creating in this chapter (Figure 2-1.1).

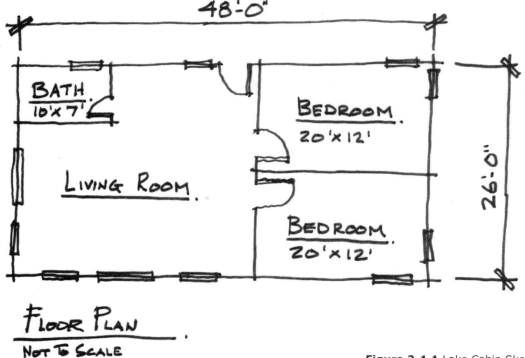

Figure 2-1.1 Lake Cabin Sketch

Exterior Walls:

1. Start a new project named **Lake Cabin**. *See Lesson 1 on creating a new project.*

2. Click on the **Wall** tool under the *Basics* tab in the **Design Bar** (Figure 2-1.2).

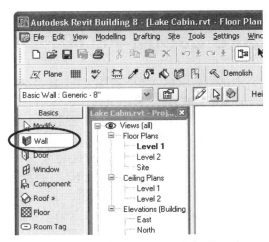
Figure 2-1.2 Wall tool

Notice that the Options Bar has changed to show options related to walls. Next you will modify those settings.

Figure 2-1.3 Options Bar

3. Modify the **Options Bar** to the following (Figure 2-1.3):
   a. *Type Selector*: Click the down-arrow and select **Basic Wall: Generic – 12"**.
   b. *Height*: Change the height from 20'-0" to **9'-0"**.
   c. *Loc Line*: Set this to **Finish Face: Exterior**.

You are now ready to draw the exterior walls.

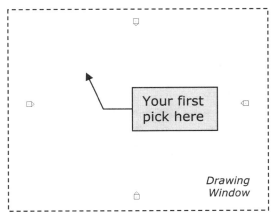

4. In the *Drawing* window, click in the upper left corner.
   TIP: *Remember to draw within the four elevation markers (image to the right).*

5. Start moving the mouse to the right. **Click** when the wall is **48'-0"** long and horizontal.

Notice as you move the mouse Revit dynamically displays a length and an angle. If you want a horizontal line, you move the mouse straight across the screen. A dashed line and a tooltip will appear when the line is snapped to the horizontal (Figure 2-1.4).

Figure 2-1.4 First wall segment

TIP: If your mouse moved a little when you clicked and the wall is not exactly 48'-0", simply click on the dimension, type 48' and then press Enter.

You are now ready to pick the first point of your second wall.

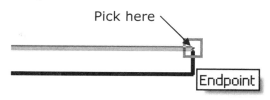

Figure 2-1.5 Second wall start point

6.   Click the right end of the first wall, making sure you snap to the outside corner of your building (Figure 2-1.5).

You may need to zoom in to pick the correct point; see Lesson 1 for zooming.

7.   Start moving your mouse straight down (south); when the dashed line and tooltip appear (indicating a vertical line), type **26'** and press **Enter**.

Typing the length allows you to accurately input a length without having to spend a lot of time setting the mouse in just the right position. However, you can still adjust the dimension after the wall is drawn.

8.   Draw the other two exterior walls.

Interior Walls:

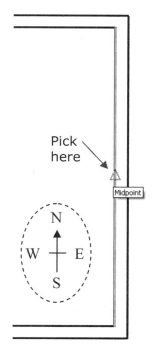

Figure 2-1.6 Interior wall start point

9.   With the Wall tool selected, modify the **Options Bar** to the following:
     a. *Type Selector*: Click the down-arrow and select **Basic Wall: Generic – 5"**.
     b. *Loc Line*: Set this to **Core Centerline**.

10. Draw a wall between bedrooms. *Snap* to the midpoint of the east wall. (Figure 2-1.6)

11. While moving the mouse to the west (left) and snapped to the horizontal plane, type **20'2 1/2**.

     NOTE: Type the length as shown; you don't need a dash or the inch symbol as they are assumed here. You do need a space before the fraction.

12. Draw the vertical wall to close off the bedrooms. Revit allows you to do this with one wall segment by selecting your points in a particular way. See **Figure 2-1.7** for a graphical description of this process. (Figure 2-1.7)

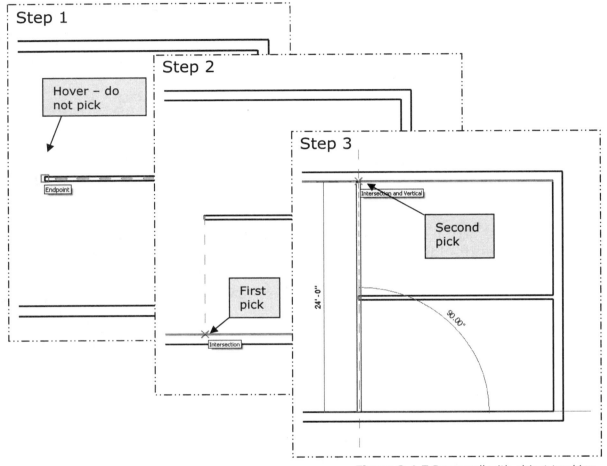

Step 1

Hover – do not pick

Endpoint

Step 2

First pick

Intersection

Step 3

Intersection and Vertical

Second pick

24'-0"

90.00°

Figure 2-1.7 Draw wall with object tracking

13. Draw the two interior walls for the bathroom to complete the interior walls (Figure 2-1.1).

14. **Save** your project (*Lake Cabin.rvt*).

TIP: You can use the **Tape Measure** tool [icon] to list the dimension between two points. This is helpful when you want to verify the clear dimension between walls and Revit is displaying a distance that is to the centerline of a wall. Simply click the icon and snap to two points and Revit will temporarily display the distance.

Exercise 2-2:
Doors

In this exercise you will add doors to your cabin floor plan.

1. Open **Lake Cabin.rvt** created in Exercise 2-1.

Placing Doors:

2. Click on the **Door** tool under the *Basics* tab in the **Design Bar** (Figure 2-2.1).

Notice that the *Options Bar* has changed to show options related to Doors. Next you will modify those settings.

The *Type Selector* indicates the door style, width and height. Clicking the down arrow to the right lists all the doors loaded into the current project.

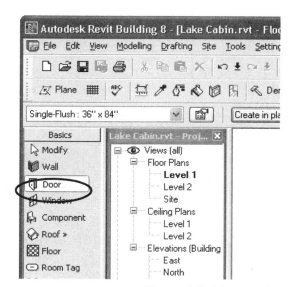

Figure 2-2.1 Door tool

The default template project that you started from has several sizes for a single flush door. Notice that there are two standard heights in the list. The 80" (6'-8") doors are the standard residential height and the 84" (7'-0") doors are the standard commercial door height.

3. Change the type selector to **Sgl Flush: 36" x 80"**.

4. Move your cursor over a wall and position the door as shown in **Figure 2-2.3**. *Notice that the swing of the door changes depending on what side of the wall your cursor is.* (Figure 2-2.3)

5. Click to place the door. Revit automatically trims the wall.

6. While the door is still selected, click on the *change swing (control arrows)* symbol to make the door swing against the wall (Figure 2-2.2).

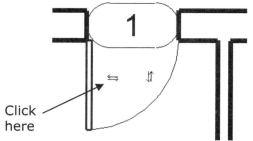

Figure 2-2.2 Changing door swing

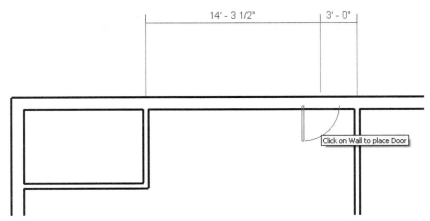

Figure 2-2.3 Dynamic graphical info during door insertion

Revit allows you to continue inserting doors until you select a different tool from the *Toolbar* or *Design Bar*.

7. Insert the doors into the bedrooms as shown in **Figure 2-1.1**. The exact position is not important in this exercise.

8. Change the *Type Selector* to **Sgl Flush: 30" x 80"**.

9. Insert a door into the bathroom.

Deleting Doors:

Next you will learn how to delete a door when needed. This process will work for most objects (i.e. walls, windows, text, etc.) in Revit.

10. Insert a door between the two bedrooms.

11. Click on the **Modify** tool on the *Design Bar*. (Figure 2-2.4)
    TIP: *Any time you press the **Esc** key Revit reverts to Modify.*

12. Click on the door you just inserted (between bedrooms) and press the **Delete** key on your keyboard.

13. **Save** your project (*Lake Cabin.rvt*).

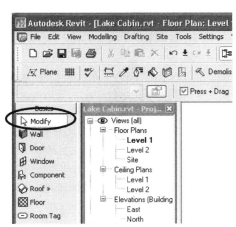

Figure 2-2.4 Modify tool

Exercise 2-3:
Windows

In this exercise you will add windows to your cabin floor plan.

1. Open **Lake Cabin.rvt** created in Exercise 2-2.

Placing Windows:

2. Click on the **Window** tool under the *Basics* tab on the **Design Bar** (Figure 2-3.1).

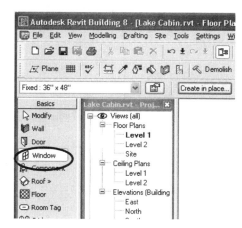

Figure 2-3.1 Window tool

Notice that the *Options Bar* has changed to show options related to Windows. Next you will modify those settings.

The *Type Selector* indicates the Window style, width and height. Clicking the down arrow to the right lists all the windows loaded in the current project.

3. Change the type selector to **Fixed: 36" x 72"**.

4. Move your cursor over a wall and place two windows as shown in **Figure 2-3.2**. *Notice that the position of the window changes depending on what side of the wall your cursor is.* (Figure 2-3.2)

5. Change the type selector to **Fixed: 24" x 72"**.

6. Place the other 4 windows in the living room area as shown in **Figure 2-1.1**. Again, in this exercise we are not concerned with the exact placement of the windows.

7. Change the type selector to **Fixed: 24" x 48"**.

8. Place the remaining 5 windows (two in each bedroom and one in the bath room) shown in **Figure 2-1.1**.

9. **Save** your project (*Lake Cabin.rvt*).

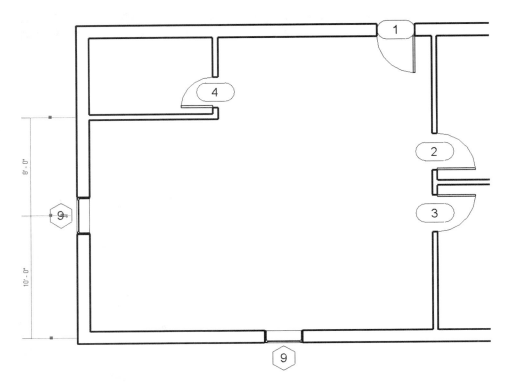

Figure 2-3.2 Two large windows

Object Snap Symbols:

By now you should be well aware of the snaps that Revit suggests as you move your cursor about the drawing window.

If you hold your cursor still for a moment while a snap symbol is displayed, a tooltip will appear on the screen. However, when you become familiar with the snap symbols you can pick sooner. (Figure 2-3.3)

The TAB key cycles through the available snaps near your cursor.

The keyboard shortcut turns off the other snaps for one pick. For example, if you type SE on the keyboard while in the Wall command, Revit will only look for an endpoint for the next pick.

Finally, typing SO (snaps off) turns all snaps off for one pick.

See Exercise 3-2 for more info.

Symbol	Position	Keyboard Shortcut
✕	Intersection	SI
☐	Endpoint	SE
△	Midpoint	SM
○	Center	SC
✕	Nearest	SN
⌐	Perpendicular	SP
Ω	Tangent	ST

Figure 2-3.3 Snap Reference Chart

Exercise 2-4:
Roof

You will now add a simple roof to your lake cabin.

1. Open **Lake Cabin.rvt** created in Exercise 2-3.

Sketching a Roof:

2. Click on the **Roof** tool under the *Basics* tab in the **Design Bar**; a fly-out menu will appear. (Figure 2-4.1)

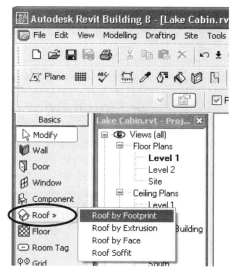

Figure 2-4.1 Roof tool

The fly-out prompts you for the method you want to use to create the roof.

3. Click **Roof by Footprint**.

Roof is on the Lowest Level warning
Revit notices that you are on Level 1 and asks you if you want to switch to another level. In our case we want to switch to Level 2. (Figure 2-4.2)

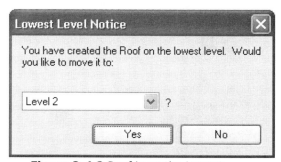

Figure 2-4.2 Roof is on the Lowest Level prompt

4. Make sure **Level 2** is selected and click **Yes** (Figure 2-4.2).

You are now on level 2 and ready to sketch the roof footprint. Notice the Level 1 walls are light grey because they are on the level below the current level (the previous step switched you to Level 2).

Also notice the *Design Bar* has temporarily been replaced with Sketch options relative to the roof (Figure 2-4.3), as with the Options Bar (2-4.4).

Figure 2-4.4
Roof sketch options

5. Pick each of the exterior walls to specify the roof footprint. *Be sure to pick the exterior side of the wall.*

6. Click **Finish Roof** on the *Design Bar*.

You should be back on Level 1. You can see this in the project browser (i.e. Level 1 is bold).

Figure 2-4.3
Roof sketch tools

7. To see the roof, click the **Default 3D View** icon. (Figure 2-4.5)

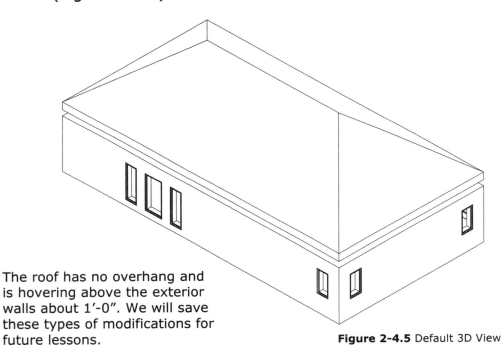

The roof has no overhang and is hovering above the exterior walls about 1'-0". We will save these types of modifications for future lessons.

Figure 2-4.5 Default 3D View

8. Click the **X** in the upper right corner of the *Drawing Window* to close the current view (3D). This will close the 3D view but not the project or the Level 1 view.

9. **Save** your project.

Exercise 2-5:
Annotation and Dimensions

Adding text is very simple in Revit. In this exercise we will add labels to each room in our lake cabin plan. We will also place two dimensions.

Placing Text:

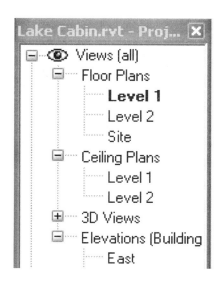

1. Open **Lake Cabin.rvt** created in Exercise 2-4.

2. Make sure your current view is **Level 1**. The word Level 1 will be bold under the heading *Floor Plans* in your *Project Browser*. If Level 1 is not current, simply double-click on the Level 1 text in the *Project Browser*. (Figure 2-5.1)

3. Select the **Text** tool under the *Basics* tab in the *Design Bar*.

Figure 2-5.1 Project Browser

Once again, notice the *Options Bar* has changed to display some text options. (Figure 2-5.2)

Figure 2-5.2 Options Bar for Text tool

The *Type Selector* indicates the text size. From this *Options Bar* you can also place text with arrow lines (leaders) and set the text alignment (i.e. Left justified, Centered or Right justified). We will not adjust these settings at this time.

4. You will now place the words "Living Room". **Click** within the living room area to place the text. (Figure 2-5.3)

5. Type **Living Room**, then click somewhere in the plan view to finish the text.

You may notice that the text seems too large. This is a good time to explain what the text height is referring to in the *Type Selector*.

The text height in the *Type Selector* refers to the size of the text on a printed piece of paper. For example, if you print your plan you should be able to place a ruler on the text and read ¼" when the text is set to ¼" in the *Type Selector*.

This can be a complicated process in other CAD programs; Revit makes it very simple. All you need to do is change the **view scale** for **Level 1** and Revit automatically adjusts the text and annotation to match that scale. Currently our *View Scale* is set to 1/8" = 1'-0"; we want the view scale to be 1/4" = 1'-0". With the view scale set to 1/8" 1'-0" our text is twice as big as it should be. Next you will change the *View Scale* for Level 1.

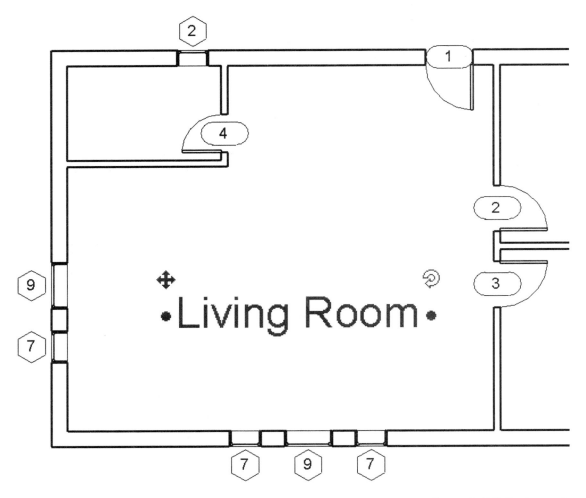

Figure 2-5.3 Placing text

6. In the lower-left corner of the Drawing Window, on the *View Control Bar*, click on the text of the listed scale (i.e. 1/8"=1'-0").

7. In the pop-up menu, which lists several standard scales, select **1/4"=1'-0"** (Figure 2-5.4).

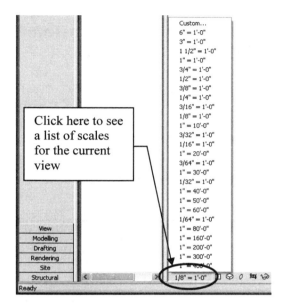

You should now notice that your text and even your door and window symbols are half the size they used to be. (Figure 2-5.5)

You should understand that this scale adjustment only affected the current view (i.e. Level 1). If you switched to Level 2 you would notice it is still set to 1/8"=1'-0". This is nice because you may, on occasion, want one plan at a larger scale to show more detail.

Figure 2-5.4 Set View Scale

8. Finally, using the **Text** tool, place a room name label in each room as shown in **Figure 2-1.1** and **Figure 2-5.5** below.

9. **Save** your project.

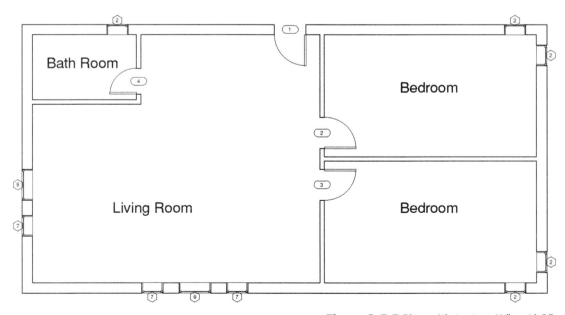

Figure 2-5.5 Plan with text at ¼" = 1'-0"

Place Dimensions:

To finish this exercise you will place two overall plan dimensions.

10. Select the **Dimension** tool under the *Basics* tab in the *Design Bar*.

11. In the *Options Bar*, change the drop-down list that says *Prefer: Wall centerlines* to **Prefer: Wall faces**. *This option will allow you to dimension to the outside face of your building, as you would normally do when dimensioning the overall footprint of your building.*

12. Place a dimension by selecting two walls and then clicking a third point to specify where the dimension line should be relative to the walls. (Figure 2-5.6)

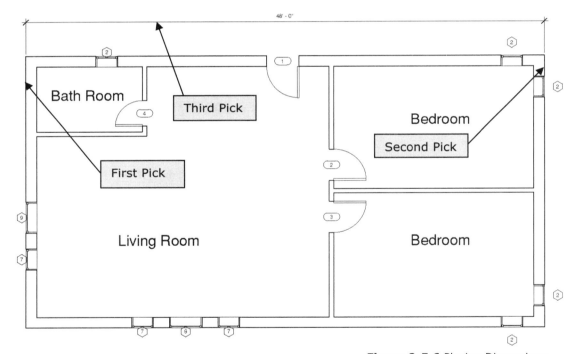

Figure 2-5.6 Placing Dimensions

13. Place one more dimension indicating the depth of the building. (Figure 2-1.1)

14. **Save** your project.

Exercise 2-6:
Printing

The last thing you need to know to round off your basic knowledge of Revit is how to print the current view.

Printing the current view:

1. Select **File → Print** from the menu bar.

2. Adjust your settings to match those shown in **Figure 2-6.1**:

    -   Select a printer from the list that you have access to;
    -   Set *Print Range* to: **Visible portion of current window**.

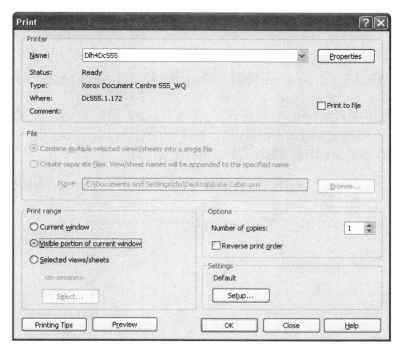

Figure 2-6.1 Print dialog

3. Click on the **Setup** button to adjust additional print settings.

4. Adjust your settings to match those shown in **Figure 2-6.2**.

    -   Set Zoom to:
        Fit to page

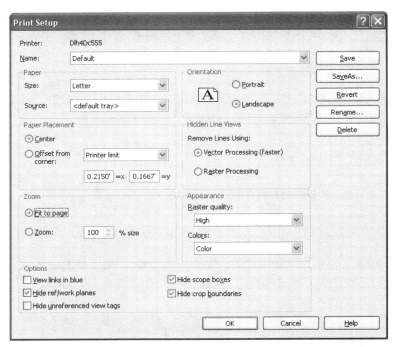

Figure 2-6.2 Print Settings dialog

5. Click **OK** to close the *Print Setup* dialog and return to *Print*.

6. Click the **Preview** button in the lower-left corner. This will save paper and time by verifying the drawing will be correctly positioned on the page (Figure 2-6.3).

7. Click the **Print** button at the top of the preview window.

8. Click **OK** to print to the selected printer.

FYI:

Notice you do not have the option to set the scale (i.e. 1/8" = 1'-0"). If you recall from our previous exercise, the scale is set in the properties for each view.

If you want a quick half-scale print, you can change the zoom factor to 50%. You could also select Fit to page *to get the largest image possible but not to scale.*

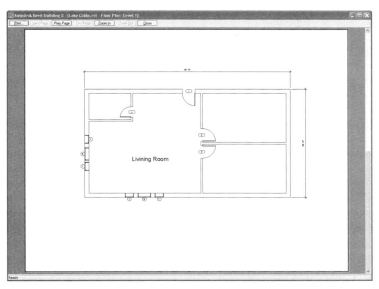

Figure 2-6.3 Print Preview

Printer versus a Plotter?
Revit can print to any printer / plotter installed on your computer.

A <u>Printer</u> is an output device that uses smaller paper (e.g. 8 ½"x11" or 11"x17"). A <u>Plotter</u> is an output device that uses larger paper; plotters typically have one or more rolls or paper ranging in size from 18" wide to 36" wide. A roll feed plotter has a built-in cutter that can, for example, cut paper from a 36" wide roll to make 24"x36" sheets.

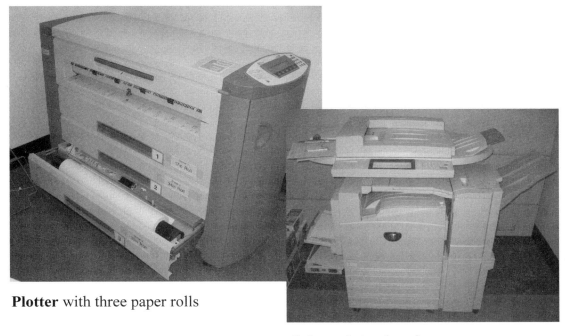

Plotter with three paper rolls

Color **printer** / copier

Self-Exam:

The following questions can be used as a way to check your knowledge of this lesson. The answers can be found at the bottom of this page.

1. The Tape Measure tool is used to dimension drawings. (T/F)

2. Revit will automatically trim the wall lines when you place a door. (T/F)

3. Snap will help you to draw faster and more accurately. (T/F)

4. A 6'-8" door is a standard door height in _____ construction.

5. While using the wall tool, the height can be quickly adjusted in the

    _____ bar.

Review Questions:

The following questions may be assigned by your instructor as a way to assess your knowledge of this section. Your instructor has the answers to the review questions.

1. The **view scale** for a drawing/view is set by clicking the scale listed on the View Control Bar. (T/F)

2. Dimensions are placed with only two clicks of the mouse. (T/F)

3. The relative size of text in a drawing is controlled by the view scale. (T/F)

4. You can quickly switch to a different view by double-clicking on that views label in the Project Browser. (T/F)

5. You cannot select which side of the wall a window is offset to. (T/F)

6. In the Print dialog box, click the _____ button to see a sample of how the printout will look.

7. The _____ tool can be used to list the distance between two walls without drawing a dimension.

8. While in the Door tool you can change the door style and size via the

    _____ _____ within the Options Bar.

Lesson 3
Overview of Linework and Modify Tools::

It may seem odd to you that, in a revolutionary 3D design program, you will begin by learning to draw and edit 2D lines and shapes. However, any 3D object requires, at a minimum, detailed information in at least two of the three dimensions; once two dimensions are defined, Revit can begin to automate much of the third dimension for you. Many building components (floors, roofs, ceilings, etc.) and features (sub-room areas that will require their own room tag) will require you to draw the perimeter using 2D lines.

Many of the edit tools that are covered in this lesson are the same tools at are used to edit the 3D building components (i.e. walls, doors, etc.).

Exercise 3-1:
Lines and Shapes

Drawing Lines

You will draw many 2D lines in Revit, typically in what is called *Sketch* mode. 2D Lines in Revit are extremely precise drawing elements. This means you can create very accurate drawings. Lines, or any drawn object, can be as precise as 8 decimal places (i.e. 24.99999999) or 1/256.

Revit is a vector based program. That means each drawn object is stored in a numerical database. When a geometry needs to be displayed on the screen, Revit reads the data from the project database to determine what to display on the screen. This means that the line will be very accurate at any scale or zoom magnification.

A raster based program, in contrast to vector based, is comprised of dots that infill a grid. The grid can vary in density, and is typically referred to as resolution (e.g. 600x800, 1600x1200, etc.). This file type is used by graphics programs that typically deal with photographs, such as Adobe Photoshop. There are two reasons this type of file is not appropriate for CAD programs:

1. A raster based line, for example, is composed of many dots on a grid (which also represents the line's width). When you zoom in (or magnify) the line, it starts to become pixilated, meaning

you actually start to see each dot in the grid. In a vector file you can "infinitely" zoom in on a line and it will never become pixilated because the program recalculates the line each time you zoom in.

2. A CAD program, such as Revit, only needs to store the starting point and end point coordinates for each wall (for example); the dots needed to draw the wall are calculated on-the-fly for the current screen resolution. Whereas a raster file has to store each dot that represents the full length and width of the line (or lines in the wall example). This can vary from a few hundred dots to several thousand dots, depending on the resolution, for the same line.

The following graphic illustrates this point:

Vector vs. Raster Lines

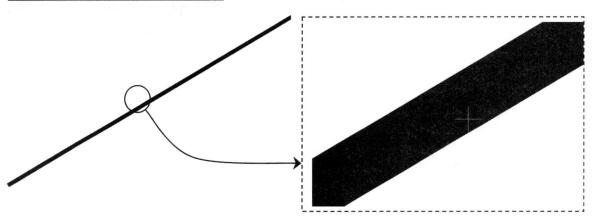

Figure 3-1.1 Vector Line Example
File Size: approx. 33kb

Figure 3-1.1a Vector Line Enlarged 1600%

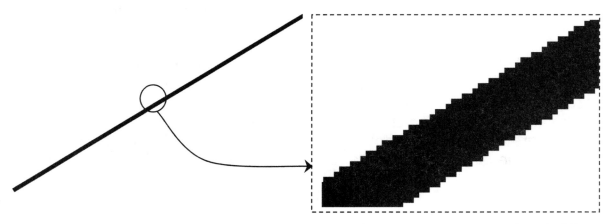

Figure 3-1.2 Raster Line Example
File Size: approx. 4.4MB

Figure 3-1.2a Raster Line Enlarged 1600%

The Detail Line Tool:

You will now study the *Detail Line* tool.

1. **Open** Revit.

Each time you start Revit, you are in the default project template. Therefore, you do not need to start a new project.

2. If Revit is already open, you should start a new project. Select **File → New → Project...**

The *Drawing Window* is set to the Level 1 floor plan view. The 2D drafting will be done in a drafting view. You will learn how to create one of these views next.

3. Select **View → New → Drafting View...**

4. In the *New Drafting View* dialog box, type **Ex3-1** for the *Name* and set the *Scale* to **3/4" = 1'-0"** by clicking the down-arrow at the right (Figure 3-1.3).

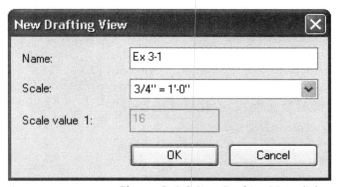

Figure 3-1.3 New Drafting View dialog

The *Project Browser* now contains a category labeled *Drafting Views*. Each *Drafting View* created will be listed here.

5. In the *Project Browser*, click the plus symbol next to the label *Drafting Views*; this will display the *Drafting Views* that exist in the current project (Figure 3-1.4).

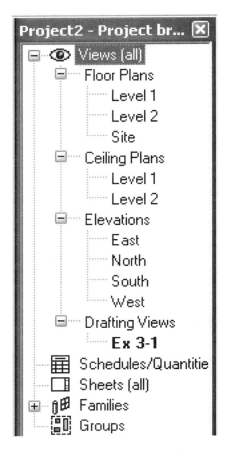

Figure 3-1.4 Project Browser:
Drafting Views

You are now ready to start looking at the Detail Line tool.

6. Select the **Drafting** tab within the *Design Bar*.

Figure 3-1.5 Drafting tab

7. On the *Drafting* tab, select **Detail Line** (Figure 3-1.5).

8. **Draw a line** from the lower left corner of the screen to the upper right corner of the screen (by simply clicking two points on the screen). ***NOTE***: *Do not drag or hold your mouse button down; just click.* (Figure 3-1.6)

After clicking your second point, you should notice the length and the angle of the line are graphically displayed; this information is temporary and will disappear when you begin the next line or cancel the tool.

You should also notice that the Detail Line tool is still active, which means you could continue to draw additional lines on the screen.

9. After clicking your second point, select the **Modify** tool from the *Drafting* tab.

Selecting *Modify* cancels the current tool and allows you to select portions of your drawing (for info / editing). Revit conveniently places the *Modify* tool on each tab in the *Project Browser*.

TIP: Pressing the **Esc** key twice reverts Revit back to Modify mode.

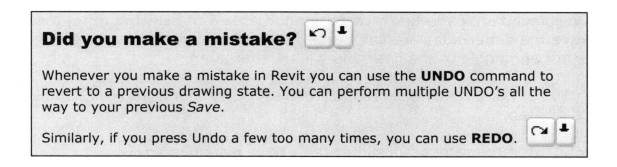

Did you make a mistake?

Whenever you make a mistake in Revit you can use the **UNDO** command to revert to a previous drawing state. You can perform multiple UNDO's all the way to your previous *Save*.

Similarly, if you press Undo a few too many times, you can use **REDO**.

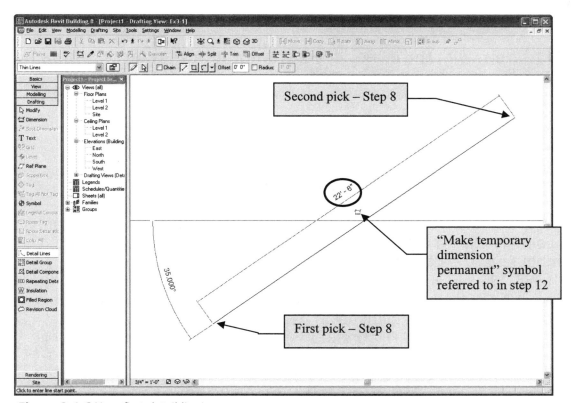

Figure 3-1.6 Your first detail line!

This constitutes your first line! However, as you are probably aware, it is not a very accurate line as far as length and angle are concerned.

Typically you need to provide information such as length and angle when drawing a line; you rarely pick arbitrary points on the screen like you did in the previous steps. Revit allows you to get close with the on-screen dimension and angle information. Most of the time you also need to accurately pick your starting point (for example, how far one line is from another or picking the exact middle point on another line).

Having said that, however, the line you just drew still has precise numbers associated with its length and angle; this information is displayed after the line is drawn. The dimension and angle information is displayed until you begin to draw another line or select another tool. While the dimensions are still visible, they can be used to modify the length and angle of the line; you will try that later.

10. If not already selected, click the **Modify** tool from the *Drafting* tab.

Notice that the temporary dimensions have disappeared.

While you are in *Modify* mode, you can select lines / objects in the current view.

11. Now, with the cursor directly over the line, **select the line** by clicking the mouse button.
    FYI: *Always click the left button unless the right button is specifically mentioned.*

Notice that the temporary dimensions have returned.

Making a Temporary Dimension Permanent:

If, at any point, you want to make a temporary dimension permanent, you simply click the "make this temporary dimension permanent" symbol near the dimension. You will try this next.

12. With the diagonal line still selected, click the "make this temporary dimension permanent" symbol near the dimension (Figure 3-1.6).

13. Select **Modify** to unselect the line.

The dimension indicating the length of the line is now permanent (Figure 3-1.7).

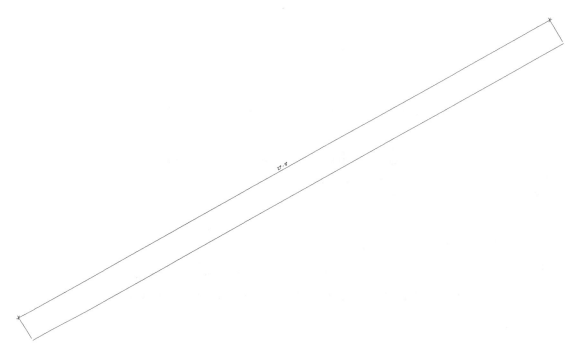

Figure 3-1.7 Temp dimension converted to permanent dimension

Changing the Length of the Line:

Currently, the line is approximately 27'–30' long (27'-9" in Figure 3-1.7). If you want to change the length of the line to 22'-6" you select the line and then change the dimension text, which in turn changes the line length. This can also be done with the temporary text; the key is that the line has to be selected. You will try this next.

    14. In *Modify* mode, select the diagonal line.

    15. With the line currently selected, click on the dimension text.

A dimension-text edit box appears directly over the dimension. Whatever you type for a dimension changes the line to match.

    16. Type **22'6** and press **Enter** (Figure 3-1.8).

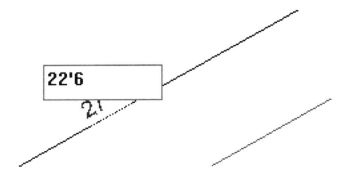

Figure 3-1.8 Editing dimension to change line length

You should have noticed the line changed length. Revit assumes that you want the line to change length equally at each end, so the midpoint does not move.

Locking Dimensions:

Revit allows you to *Lock* a dimension which prevents the dimension / line length from changing. You will investigate this now to help avoid problems later.

17. Make sure the line is *not* selected (press Esc or Modify).

18. <u>Select the dimension</u> (not the line) and note the following about the selected dimension (Figure 3-1.9).

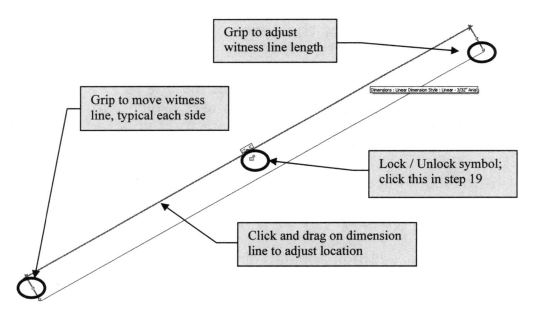

Grip to adjust witness line length

Dimensions : Linear Dimension Style : Linear - 3/32" Arial

Grip to move witness line, typical each side

Lock / Unlock symbol; click this in step 19

Click and drag on dimension line to adjust location

Figure 3-1.9 Selected dimension

19. Click on the **Lock / Unlock** (padlock) symbol to "Lock" the dimension (Figure 3-1.9).

The dimension is now locked. Again, this means that you cannot change the dimension OR the line. The dimension is attached to the line in such a way that changing one affects the other. Next you will see what happens when you try to modify the dimension while it is locked.

Any time the line or dimension are selected you will see the Lock / Unlock symbol in the locked position. *FYI: Clicking it anytime it is visible will toggle the setting between locked and unlocked.*

20. Select **Modify** to unselect the *Dimension*.

21. <u>Select the line</u>.

22. <u>Click on the Dimension Text</u> and attempt to modify the line length by typing **21'** and then press **Enter**.

You should now see the following error message (Figure 3-1.10), which indicates an error that cannot be ignored.

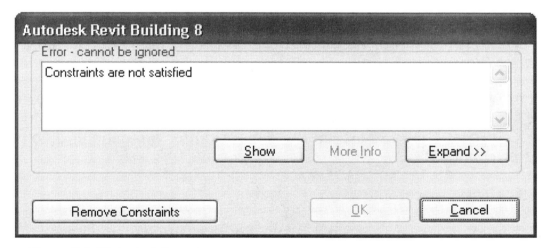

Figure 3-1.10 Locked dimension error

As a new Revit user, you will likely see messages like this often until you understand how Revit expects objects to work / relate to each other. This is helpful, because it identifies many items that deviate from your design intent or are not buildable in the real world.

If you are not sure what the constraint is, you may be able to figure it out by selecting the *Expand >>* button.

23. Click the **Expand >>** button.

24. Click the "plus" symbols to expand the view.

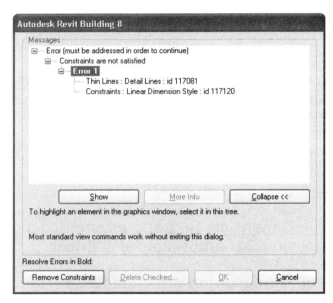

Figure 3-1.11 Locked dimension error – expanded

The *Error 1* section indicates that a *Detail Line* is being constrained by a *Linear Dimension Style*.

At this point you can press *Cancel* to ignore the attempted change or click *Remove Constraints* to unlock the dimension (in this case). In this case you recall that you locked the dimension for a reason and do not want to change it:

25. Click **Cancel**.

Even though your current view is showing a relatively small area, i.e. *Drafting View,* it is actually an infinite drawing board.

In architectural CAD drafting, everything is drawn "real-world" size (or full-scale) ALWAYS! If you are drawing a building, you might draw an exterior wall that is 600'-0" long. If you are drawing a window detail, you might draw a line that is 8" long. In either case you are entering that line's actual length.

You could, of course, have an 8" line and a 600'-0" line in the same drawing. Either line would be difficult to see at the current drawing magnification (i.e. approx 22' x 16' area; also recall that your diagonal line is 22'-6" long). So you would have to zoom in to see the 8" line and zoom out to see the 600'-0" line. You will try this next:

Draw an 8" Line:

The next steps will walk you through drawing an 8" horizontal line. Revit provides more than one way to do this. You will try one of them now.

26. Select **Detail Lines** from the Drafting tab.

27. Pick a point somewhere in the upper left corner of the *Drawing Window*.

28. Start moving the mouse towards the right, and generally horizontal until you see a dashed reference line (extending in each direction of your line) appear on the screen as in **Figure 3-1.12**.

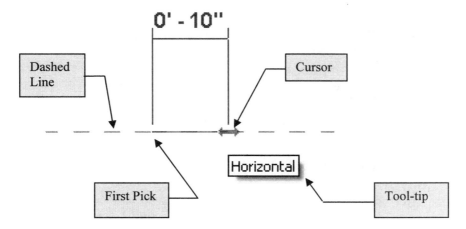

Figure 3-1.12   Drawing a line with the help of Revit
TIP: You will see a dashed horizontal line and a tooltip when the line is horizontal.

As you move the mouse left and right, you will notice Revit displays various dimensions that allow you to quickly draw your line at these oft used increments. Read the sidebar article in the next exercise to learn more about how to control the increments shown.

29. With the dashed line and *Tool-tip* visible on the screen, take your hand off the mouse (so you don't accidentally move it), and type: **8"** and then press **Enter**.

TIP: *Remember, you have to type the inch symbol; Revit always assumes you mean feet unless you specify otherwise. A future lesson will review the input options in a little more detail.*

You have just drawn a line with a precise length and angle!

30. Use the **Tape Measure** tool to verify it was drawn correctly.

31. Use the **Zoom In Region** tool to enlarge the view of the 8" line.

32. Now use the **Zoom All to Fit** or **Previous Scroll/Zoom** tools so that both lines are visible again.

Draw a 600' Line:

33. Select the **Detail Lines** tool and pick a point in the lower right corner of the *Drawing Window*.

34. Move the cursor (i.e. mouse) straight up from your first point so that the line snaps to the vertical and the dashed line appears (Figure 3-1.13).

35. With the dashed line and *tool-tip* visible on the screen, take your hand off the mouse (so you don't accidentally move it), and type: **600** and then press **Enter**.

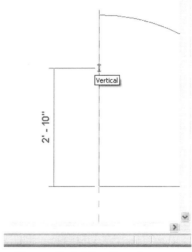

Figure 3-1.13 Drawing another Detail Line; 600'-0" vertical line – lower right

TIP #1: *Notice this time you did not have to type the foot symbol (').*

TIP #2: *You do not need to get the temporary dimension close to reading 600'; the temporary dimension is ignored when you type in a value.*

36. Press the **Esc** key twice to exit the *Drafting Line* tool and return to the *Modify* tool.

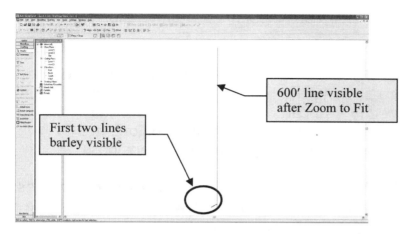

Because the visible portion of the drawing area is only about 16' tall, you obviously would not expect to see a 600' line. You need the change the drawings magnification (i.e. zoom out) to see it.

Figure 3-1.14   Detail view with three lines

37. Use **Zoom All to Fit** to see the entire drawing (Figure 3-1.14)

Drawing Shapes:

Revit provides several tools to draw common shapes like square/rectangles, circles, ellipses. These tools can be found on the Options Bar while the Detail Line tool is active. You will take a look at Rectangle and Circle now.

Rectangle:

38. Use **Previous Scroll/Zoom** (or Zoom Region) to get back to the original view.

39. Select the **Detail Lines** tool from the *Drafting Tab*.

Notice the *Options Bar* has changed to show various options related to Detail Lines *(bar partially shown below)*.

40. Select the **Rectangle** tool from the *Options Bar*.
    (**TIP**: Hover cursor over icon to see the tooltip until you learn the icons.)

41. Pick the "**first corner point**" somewhere near the middle-bottom of the *Drawing Window* (Figure 3-1.15)

Notice the temporary dimensions are displaying a dimension for both the width and height.

At this point you can pick a point on the screen, trying to get the dimensions correct before clicking the mouse button. If you do not get the rectangle drawn correctly you can click on each dimension-text and change the dimension while the temporary dimensions are disabled; see the next two steps for the rectangle dimensions.

42. Draw a 2'-8" x 4'-4" rectangle using the temporary dimensions displayed on the screen; if you do not draw it correctly do the following:

  a. Click the dimension-text for the horizontal line and then type **2'-8"** and then press **Enter**.

  b. Type **4'-4"** and then press **Enter** for the vertical dimension.

That's it; you just created your first rectangle!

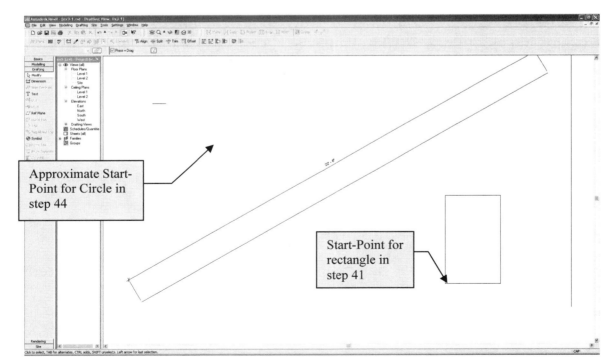

Figure 3-1.15   Drawing with Rectangle added

Circle:

43. With the *Detail Lines* tool active, select **Circle** from the *Options Bar*. **TIP**: *Click the down arrow.* (Figure 3-1.16)

Figure 3-1.16   Detail Line Options

44. You are now prompted to pick the center point for the circle; pick a point approximately as shown in Figure 3-1.15.

You should now see a dynamic circle and temporary dimension attached to your cursor (which allows you to visually specify the circle's size. Move your mouse around to see that you could arbitrarily pick a point on the screen to create a quick circle if needed, then proceed to step 45 where you will draw a circle with a radius of 1'-6 5/8".

45. Type **1'-6 5/8"** and then press **Enter** (Figure 3-1.17).

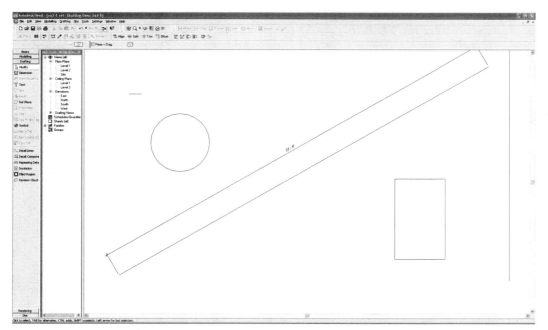

Figure 3-1.17   Drawing with Circle added

46. **Save** your drawing as **ex3-1.rvt**.

Exercise 3-2:
Snaps

Snaps are tools that allow you to accurately pick a point on an object. For example: when drawing a line, you can use *Object Snap* to select, as the start-point, the endpoint or midpoint of another line (or wall).

This feature is absolutely critical to drawing accurate technical drawings. Using this feature allows you to be confident you are creating perfect intersections, corners, etc. (Figure 3-2.1)

Object Snaps Options:
You can use *Object Snaps* in one of two ways.

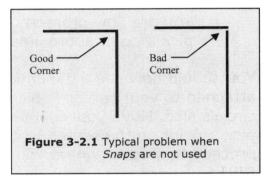

Figure 3-2.1 Typical problem when *Snaps* are not used

- o "Normal" mode
- o "Temporary Override" mode

"Normal" *Osnap* mode is a feature that constantly scans the area near your cursor when Revit is looking for user input. You can configure which types of points to look for.
Note: *The term "normal" is not a Revit term; rather it is simply a description used in this book to differentiate this portion of the Snaps feature from the Temporary Override portion discussed next.*

Using a "Temporary Override" *Object Snap* for individual pick-points allows you to quickly select a particular point on an object. This option will temporarily disable the other *Object Snap* settings, which means you tell Revit to just look for the endpoint on an object (for the next pick-point only) rather than the three or four types being scanned for by the "normal" *Object Snaps* feature.

Overview of the Snaps Dialog Box:

Revit provides a *Snaps* dialog box, where you specify which Object Snaps you want enabled. You will take a look at it next.

1. From the *Settings* pull-down menu, select **Snaps...**

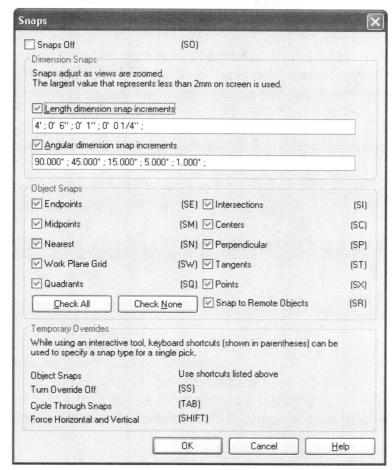

Figure 3-2.2 Snaps dialog box; default settings

You should now see the *Snaps* dialog box (Figure 3-2.2). Take a minute to study the various settings that you can control with this dialog box. Notice the check box at the top that allows you to turn this feature off completely.

Dimensions Snaps

This controls the automatic dimensions that Revit suggests when you are drawing. Following the given format you could add or remove increments.

Object Snaps

This controls the "normal" *Object Snaps* that Revit scans for while you are drawing. Un-checking an item tells Revit not to scan for it anymore. Currently, all Object Snaps are enabled (Figure 3-2.2).

<u>*Temporary Overrides*</u>

This area is really just information on how to use Temporary Overrides.

While drawing, you will occasionally want to tell Revit you only want to *Snap* to an Endpoint for the next pick. Instead of opening the *Snaps* dialog and un-checking every *Object Snap* except for Endpoint, you can specify a *Temporary Override* for the next pick. You do this by typing the two letters (in parentheses) in the *Object Snaps* area of the *Snaps* dialog (Figure 3-2.2).

2. Make sure the *Snaps Off* option is <u>not</u> checked in the upper-left corner of the dialog box (this turns Snaps completely off).

3. Click on the **Check All** button to make sure all *Snaps* are checked.

4. Click **OK** to close the *Snaps* dialog box.

Snap Symbols:

By now you should be well aware of the Object Snap Symbols that Revit displays as you move your cursor about the drawing window (while you are in a tool like *Wall* and Revit is awaiting your input or pick-point).

If you hold your cursor still for a moment while a snap symbol is displayed, a tooltip will appear on the screen. However, when you become familiar with the snap symbols you can pick sooner (rather than waiting for the tooltip to display). (Figure 3-2.3)

The **Tab** key cycles through the available snaps near your cursor.

Finally, typing SO turns all snaps off for one pick.

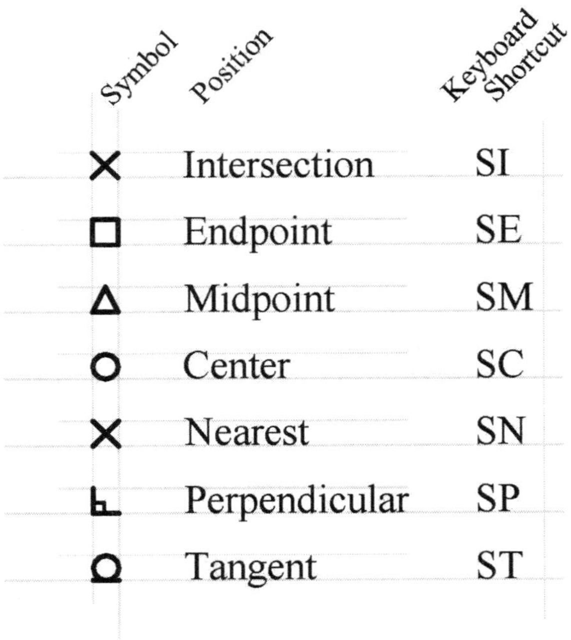

Symbol	Position	Keyboard Shortcut
✕	Intersection	SI
☐	Endpoint	SE
△	Midpoint	SM
○	Center	SC
✕	Nearest	SN
⌐	Perpendicular	SP
Ω	Tangent	ST

Figure 3-2.3 Snap symbols

Setting Object Snaps:

You can set Revit to have just one or all Snaps running at the same time. Let's say you have Endpoint and Midpoint set to be running. While using the *Wall* tool, you move your cursor near an existing *Wall*. When the cursor is near the end of the *Wall* you will see the Endpoint symbol show up; when you move the cursor towards the middle of the line you will see the Midpoint symbol show up.

The next step shows you how to tell Revit which Object Snaps you want it to look for.

FYI...

The *Snaps* shown in Figure 3-2.2 are for Revit in general, not just the current project.

This is convenient; you don't have to adjust to your favorite settings for each drawing (existing or new).

First you need to **Open** the project from the previous Lesson.

1. **Open** project **ex3-1.rvt** (if not currently open).

2. Next do a **Save-As** and name the project **ex3-2.rvt**.

3. As discussed previously, open the **Snaps** dialog box.

4. Make sure only the following *Snaps* are checked:
   a. Endpoint
   b. Midpoint
   c. Center
   d. Intersection
   e. Perpendicular

5. Click **OK** to Close the dialog box.

FOR MORE INFORMATION...

For more on using Snaps, search Revit's Help System for **Snaps**.

Then double-click the **Snap Points** or any other items found by the search.

Now that you have the *Snaps* set, you will give this feature a try.

6. Using the **Detail Lines** tool, move your cursor to the lower-left portion of the diagonal line (Figure 3-2.4).

7. Hover the cursor over the line's endpoint (without picking). When you see the *Endpoint* symbol you can click to select that point.

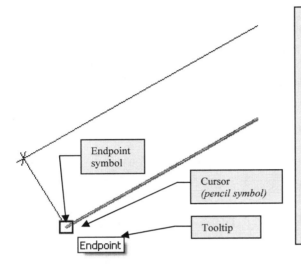

> ### *Just so you know...*
> It is important that you see the *OSNAP* symbol before clicking. Also, once you see the symbol you should be careful not to move the mouse too much.
>
> These steps will help to ensure accurate corners.

Figure 3-2.4 Endpoint SNAP symbol visible

While still in the Drafting Lines tool you will draw lines using *Snaps* to draft accurately.

8. Draw the additional lines shown in **Figure 3-2.5** using the appropriate *Object Snap* (changing the selected Snaps as required to select the required points).

    a. *TIP #1*: When using the *Drafting Lines* command you can draw several line segments without having to end the Line tool and then restart it for the next line segment. To do this you must have the *Chain* option checked on the *Options Bar*. After picking the start and end points for a line, the end point automatically becomes the first point for the next line segment; this continues until you finish the Line tool (i.e. right-click/Cancel or Esc twice).

    b. *TIP #2*: At any point, while the Line tool is active, you can open the SNAP dialog box and adjust its settings. This will not cancel the line command.

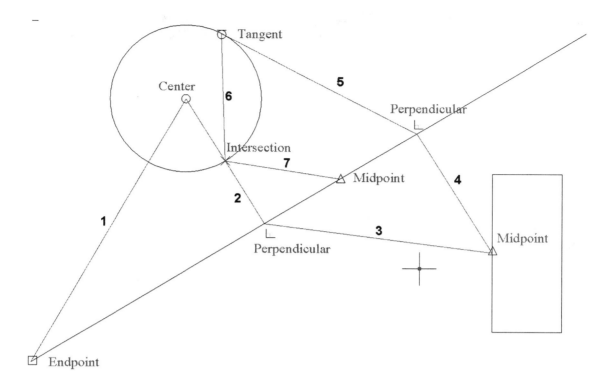

Figure 3-2.5 Lines to draw

9. **Save** your project.

This is a residential example?

OK, this is not really an architectural example yet. However, the point here is to focus on the fundamental concepts and not architecture just yet.

Exercise 3-3:
Edit Tools

The Edit Tools are used often in Revit. Much time is spent tweaking designs and making code / client related revisions.

Example: you use the Design tools (e.g. Walls, Doors, Windows) to initially draw the project. Then you use the Edit tools to *Move* a wall so a room becomes larger, *Mirror* a cabinet so it faces in a different direction, and *Rotate* the furniture per the owner's instructions.

You will usually access the various Edit tools from the Standard *or* Edit Toolbars; you can also access them from the Edit Menu shown in Figure 3-3.1 (You can probably visualize what most of the commands do by their names). The two letters shown to the right of a tool, in the menu, are its keyboard shortcut; pressing those two keys, one at a time, activates that tool.

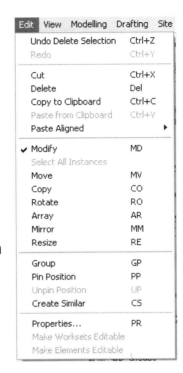

Figure 3-3.1 Edit menu

In this exercise you will get a brief overview of a few of the Edit tools, manipulating the tangled web of lines you have previously drawn.

1. **Open** project **ex3-2.rvt** from the previous lesson.

2. **Save-As ex3-3.rvt**.
    FYI: You will notice, in this book, that instructions or tools that have already been covered will have less "step-by-step" instruction.

Delete Tool:

It is no surprise that the *Delete* tool is a necessity; things change and mistakes are made. You can *Delete* one object at a time or several. Deleting objects is very easy; you select the Object (or Objects), and then pick the *Delete* icon. You will try this on two lines in your drawing.

3. While holding the Ctrl key, use the *Modify* tool to select the lines identified in **Figure 3-3.2**.
   TIP: See the section below on *Selecting Entities*.

4. Select **Delete**. ✕

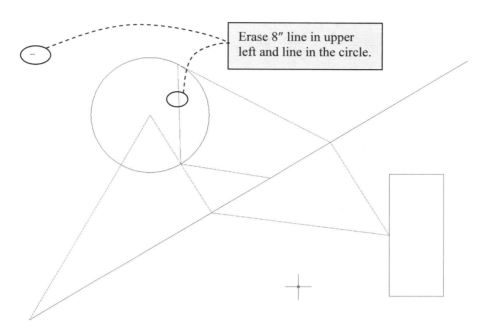

Erase 8″ line in upper left and line in the circle.

Figure 3-3.2 Lines to be erased

The lines are now deleted from the project.

Selecting objects:

At this time we will digress and take a quick look at the various techniques for selecting entities in Revit. Most tools work the same when it comes to selecting objects.

When selecting entities, you have two primary ways to select them:
- o Individually select entities one at a time
- o Select several entities at a time with a window

Continued on next page

You can use one, or a combination of both, methods to select objects (when using the Copy tool for example).

Individual Selections:
When prompted to select entities (to copy or delete, for example), you simple move the cursor over the object and click; hold the Ctrl key to select multiple objects – then you typically select the tool you wish to use on the selected items.

Window Selections:
Similarly, you can pick a window around several objects to select them all at once. To select a window, rather than selecting an object as previously described, you select one corner of the window you wish to define (that is, you pick a point in "space"). As you move the mouse you will see a rectangle on the screen that represents the windowed area you are selecting. When the window encompasses the objects you wish to select, click the mouse.

You actually have two types of windows you can select. One is called a **Window** and the other is called a **Crossing Window**.

Window:
This option allows you to select only the objects that are completely within the window. Any lines that extend out of the window are not selected.

Crossing Window:
This option allows you to select all the entities that are completely within the window and any that extend outside the window.

Using Window versus Crossing Window:
To select a *Window* you simply pick and drag from left to right to form a rectangle (Figure 3-3.3a).

Conversely, to select a Crossing Window, you pick and drag to define two diagonal points of the window from right to left (Figure 3-3.3b).

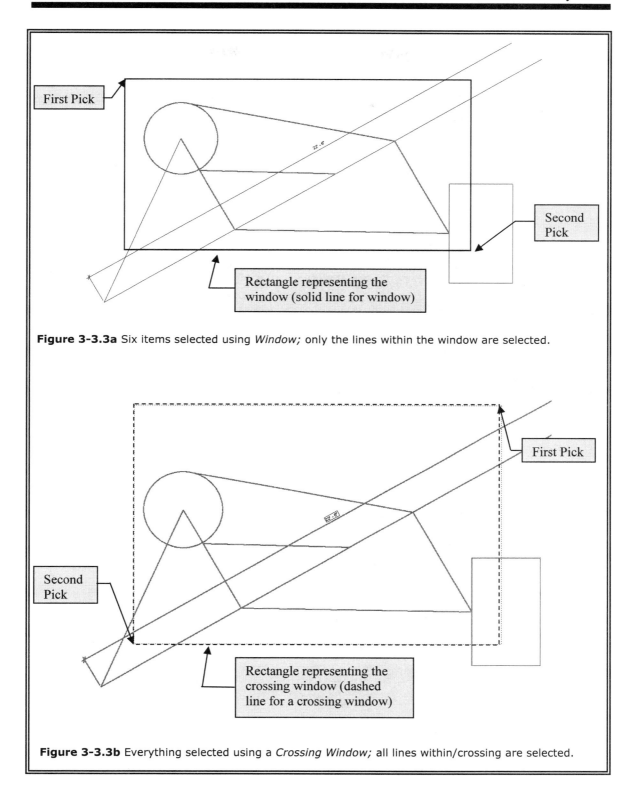

Figure 3-3.3a Six items selected using *Window;* only the lines within the window are selected.

Figure 3-3.3b Everything selected using a *Crossing Window;* all lines within/crossing are selected.

Copy tool:

The Copy tool allows you to accurately duplicate an object(s). You select the items you want to Copy and then pick two points that represent an imaginary vector (which provides both length and angle) defining the path used to copy the object to; you can also type in the length and angle if there are no convenient points to pick in the drawing. You will try both methods next.

5. **Select the circle**.

6. Select **Copy**.   [→] Copy

Notice the prompt at the bottom: click to enter move start point.

7. Pick the **Center** of the **Circle** (Figure 3-3.4).
   FYI: *You actually have three different Snaps you can use here: Center, Endpoint and Intersection. All occur at the exact same point.*

Notice the prompt at the bottom: click to enter move end point.

8. Pick the Endpoint of the angled line in the lower left corner (Figure 3-3.4).
   FYI: *If you want to make several copies of the circle, select Multiple on the Options Bar.*

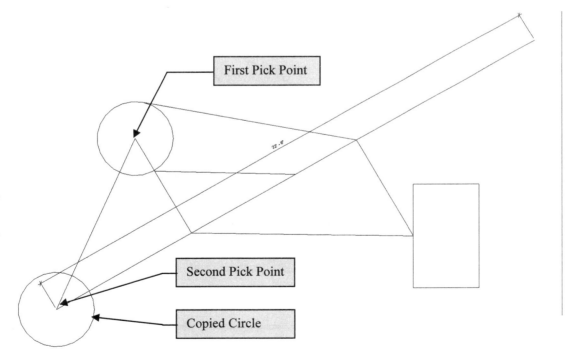

Figure 3-3.4 Copied circle; also indicates points selected

9. **Select the rectangle** *(all four lines)*.

10. Select **Copy**.

11. Pick an arbitrary point on the screen. *(In this scenario it makes no difference where you pick; you will see why in a moment.)*

At this point you will move the mouse in the direction you want to copy the rectangle (until the correct angle is displayed) and then type in the distance, rather than picking a second point on the screen.

12. Move the mouse towards the upper right until 45 degrees displays and then type 6' and then press enter. (Figure 3-3.5)

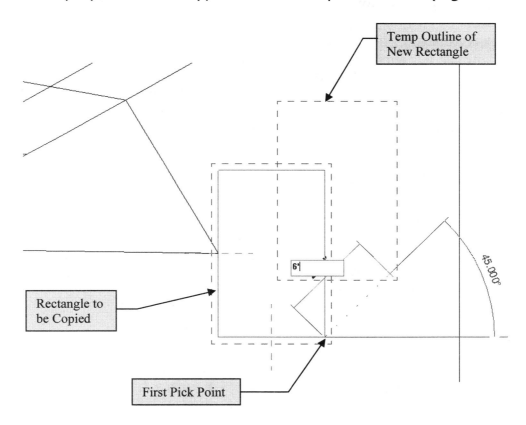

Figure 3-3.5 copying rectangle; angle set visually and distance typed in

Note: Your drawing may look a little different than this one because your drawing is not totally to scale.

Move tool:

The Move tool works exactly like the Copy tool except, of course, you move the objects rather than copy them. Given the similarity to the previous tool covered, the Move tool will not be covered here. You are encouraged to try it yourself.

Rotate tool:

With the Rotate tool, you can arbitrarily or accurately rotate one or more objects in your drawing. When you need to Rotate accurately, you can pick points that define the angle (assuming points exist in the drawing to pick) or you can type a specific angle.

Rotate involves the following steps:
- Select object(s) to be rotated
- Select Rotate
- Determine if "center of rotation" symbol is where you want it
    o If you want to move it you simply drag it
- Select a point to define a reference line and begin rotation
- Pick a second point *(using one of the following methods)*
    o by selecting other objects or using the graphic angle display
    o you can type an angle and press enter

13. **Select the new rectangle** you just copied *(all four lines)*.

14. Select **Rotate**.

Notice that the "Center of Rotation" symbol, by default, is located in the center of the selected objects. This is the point about which the rotation will occur. See Figure 3-3.6.

You will not change the "Center of Rotation" at this time.

15. Pick a point directly to the right of the "Center of Rotation" symbol; this tells Revit you want to rotate the rectangle relative to the horizontal plane (Figure 3-3.6).

Now, as you move your mouse up or down, you will see the temporary angle dimension displayed. You can move the mouse until the desired angle is displayed and then click to complete the rotation, or you can type the desired angle and then press Enter. If you recall, the Snaps dialog box controls the increments of the angles that Revit displays (they are all whole numbers), so if you need to rotate something 22.45 degrees, you must type it as Revit will never display that number as an option.

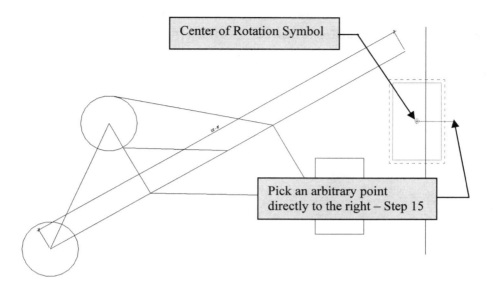

Figure 3-3.6 Rotate tool – First Step; Select lines and then click Rotate

16. Move the mouse up and to the left until 90 degrees is displayed and then click to complete the rotation. (Figure 3-3.7)

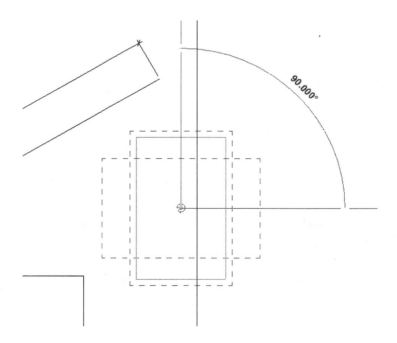

Figure 3-3.7 Rotate tool – Last Step; Select angle visually or type angle

Thus, the previous steps just rotated the rectangle 90 degrees counter-clockwise about its center point. (Figure 3-3.8)

17. Select the **Undo** icon.

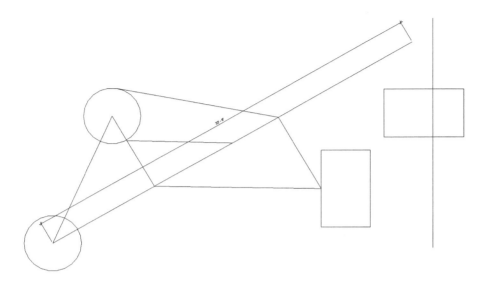

Figure 3-3.8 Rotate tool – Rectangle rotated about its center point

Now you will do the same thing, except with a different angle and "Center of Rotation".

18. **Select the rectangle** and then pick **Rotate**.

19. Click and drag the "Center or Rotation" symbol to the left and *Snap* to the midpoint of the vertical line. (Figure 3-3.9)

20. Pick a point to the right (creating a horizontal reference line).

21. Start moving your mouse downward, type **22.5**, and then press **Enter** (Figure 3-3.10).

The rectangle is now rotated 22.5 degrees in the clockwise direction (Figure 3-3.11).

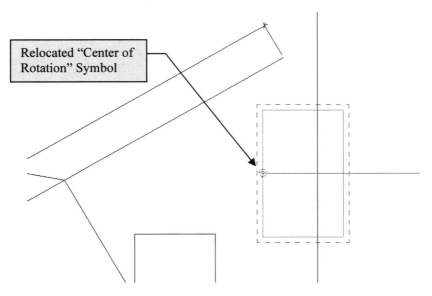

Relocated "Center of Rotation" Symbol

Figure 3-3.9 Rotate tool; relocated "center of rotation"

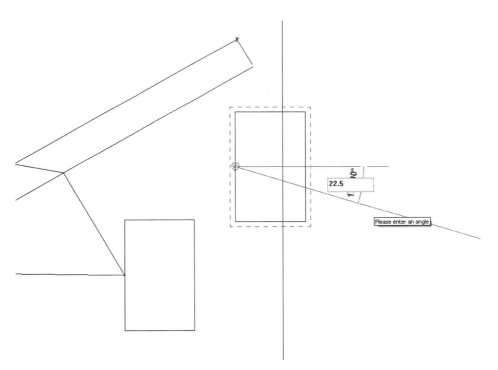

22.5

Please enter an angle

Figure 3-3.10 Rotate tool; typing in exact angle

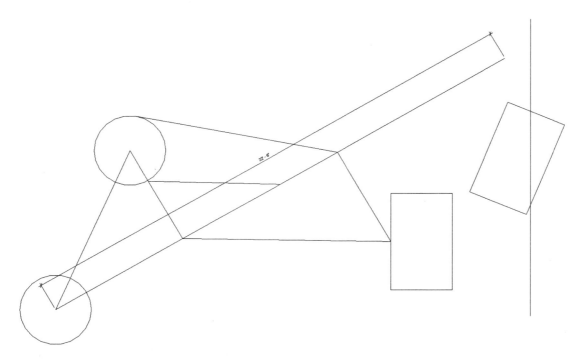

Figure 3-3.11 Rotate tool; Rectangle rotated 22.5 degrees

Resize tool:

The Resize tool has steps similar to the Rotate tool. First you select what you want to resize (or scale), specify a scale factor and then you pick the base point. The scale factor can be provided by picks on the screen (graphical) or entering a numerical scale factor (e.g. 2 or .5, where 2 would be twice the original size and .5 would be half the original size).

Next, you will use the Resize tool to adjust the size of the circle near the bottom.

Before you resize the circle, you should use the **Modify** tool to note the **diameter** of the circle (select the circle and view its Temp dimensions). After resizing the circle, you will refer back to the Temp dimensions to note the change. This step is meant to teach you how to verify the accuracy and dimensions of entities in Revit.

22. **Select the circle** (the one at the bottom).

23. Select the **Resize** icon.

On the *Options Bar* you will specify a numeric scale factor of .5 to resize the circle to half its original size.

24. Click **Numerical** and then type **.5** in the text box.
(See the Options Bar above.)

You are now prompted: **Click to enter origin** on the *Status Bar*. This is the base point about which the circle will be scaled (see examples in Figure 3-3.13 on the next page).

Now use the **Modify** tool to note the change in the size of the circle, from 1'-6 5/8" to 9 5/16" radius. A Scale Factor of .5 reduces the entities to half their original scale. (Figure 3-3.12)

> ### Selecting the correct center of rotation *(base point)*:
> You need to select the appropriate Center of Rotation (or Base Point), for both the Resize and Rotate commands, to get the results desired. A few examples are shown in Figure 3-3.13. The dashed line indicates the original position of the entity being modified. The black dot indicates the base point selected.

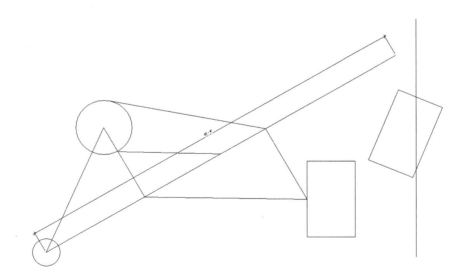

Figure 3-3.12 Resize circle; notice it is half the size of the other circle

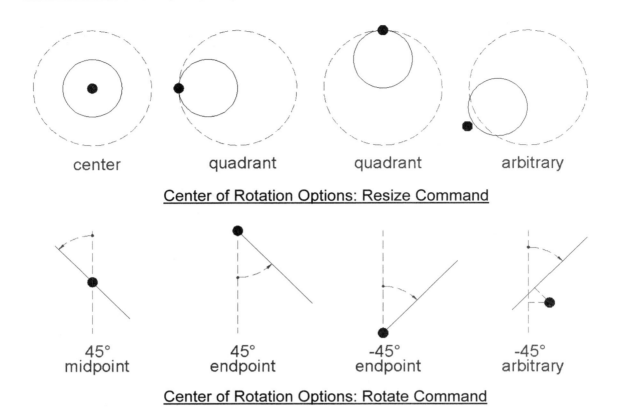

center   quadrant   quadrant   arbitrary

Center of Rotation Options: Resize Command

45°
midpoint   45°
endpoint   -45°
endpoint   -45°
arbitrary

Center of Rotation Options: Rotate Command

Figure 3-3.13 Center of Rotation Options: the various results

This will conclude the brief tour of the Edit tools. Don't forget, even though you are editing 2D lines, the edit tools work the same way on 3D objects such as walls, doors, roofs, etc.

As you surely noticed, the Edit menu and toolbar have a few other tools that have not been investigated yet. Many of these will be cover later in this book.

25. **Save** your project.
(Your project should already be named *ex3-3.rvt* per step 2 on page 3-22.)

Carter Residence

Images courtesy of LHB
www.LHBcorp.com

Exercise 3-4:
Annotations

Annotations (text, notes) allow the designer / drafter to accurately describe the drawing. You will take a quick look at this feature now.

Annotations:

Adding annotations to a drawing can be as essential as the drawing itself. Often the notes (i.e. annotations) describe something about a part of the drawing that would be difficult to discern from the drawing alone.

For example: a wall framing drawing showing a bolt may not indicate, graphically, how many bolts are needed or at what spacing. The note might say *"5/8" anchor bolt at 24" O.C."*

Next you will add text to your drawing.

1. **Open** project **ex3-3.rvt** from the previous lesson.

2. **Save-As ex3-4.rvt**.

3. Use **Zoom**, if required, to see the entire drawing (except for the 600' line which can run off the screen).

4. From the Drafting tab (or the Basic tab) select the **Text** tool. *(Notice that some tools are repeated on additional tabs for convenience.)*   T Text

From the Drafting pull-down menu you can see that the keyboard shortcut for the *Text* tool is TX. Thus, at any time you can type T and then X to activate the Text tool; if you have another command active, Revit will cancel it.

Experienced / efficient designers use a combination of keyboard shortcuts and mouse clicks to save time. For example, while your left hand is typing TX, your right hand can be moving toward the location where you want to insert the text.

Notice the current *Prompt* on the *Status Line* at the bottom of the screen; you are asked to "Click to start text or click and drag to create text wrapping". By clicking on the screen, you create one line of text (starting at the point picked) and press **Enter** to create additional lines. By clicking and dragging you specify a window, primarily for the width, which causes Revit to automatically move text to the next line (i.e. text wrapping) when the text no longer fits within the window. (Figure 3-4.1)

Figure 3-4.1 Status line while Text tool is active

You should also notice the various text options, available on the Options Bar, while the Text tool is active (Figure 3-4.2). You have two text heights (loaded from the template file you started with), text justification icons (Left, Center, Right), and the option to attach leaders (i.e. pointing arrows) to your drawings. The Text tool will be covered in more detail later in this book.

FYI: The text heights shown (i.e. 1/4", 3/32") are heights the text will be when printed on paper, regardless of drawing scale.

Figure 3-4.2 Options Bar with the Text tool active

5. Pick a point in the upper left portion of the view as shown in **Figure 3-4.3**.

6. Type **Learning Revit Building 8 is fun!**

7. Click anywhere in the view (except on the active text).

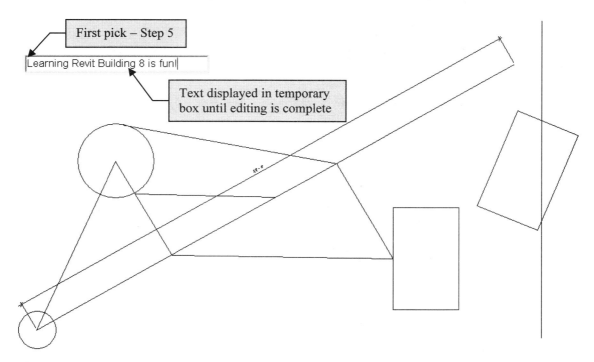

Figure 3-4.3 Text tool: Picking first point and typing text.

As soon as you complete the text edit, the text will still be selected and the "Move" and "Rotate" symbols will be displayed near the text; this allows you to more quickly *Move* or *Rotate* the text after it is typed (Figure 3-4.4).

Figure 3-4.4 Text tool; text typed with Move / Rotate symbols showing.

8.   Finish the Text tool completely by clicking away from the text.
     TIP: *Notice that the text is no longer selected and the symbols are gone.*

Your text should look similar to Figure 3-4.5. Text can be Moved and Rotated with the tools on the *Modify Toolbar*.

9.   Print your drawing (refer back to page 2-15 for basic printing information). Your print should fit the page and look similar to Figure 3-4.5.

10. **Save** your project.

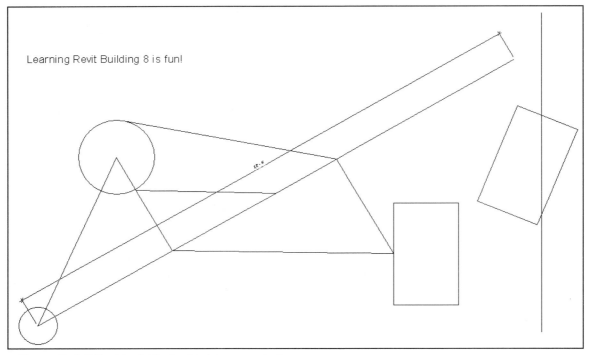

Learning Revit Building 8 is fun!

Figure 3-4.5 Text tool: Text added to current view

Carter Residence

Images courtesy of LHB
www.LHBcorp.com

Self-Exam:

The following questions can be used as a way to check your knowledge of this lesson. The answers can be found at the bottom of this page.

1. Revit is only accurate to 3 decimal places. (T/F)

2. The diamond shaped Snap symbol represents the Midpoint snap. (T/F)

3. You can move or rotate text via symbols when text is selected. (T/F)

4. Use the _____ tool to duplicate objects.

5. When selecting objects, use the _____ _____ to select all the objects in the selection window and all objects that extend through it.

Review Questions:

The following questions may be assigned by your instructor as a way to assess your knowledge of this section. Your instructor has the answers to the review questions.

1. Revit is a raster based program. (T/F)

2. The "Center of Rotation" you select for the Rotate and Resize commands are not important. (T/F)

3. Entering 16 for a distance, actually means 16'-0" to Revit. (T/F)

4. Use the *Detail Lines* tool, with _____ checked on the *Options Bar*, to create squares.

5. You can change the height of the text from the *Type Selector*. (T/F)

6. Pressing the _____ key cycles you through the snap options.

7. Where do the two predefined text heights come from? _____

8. Specifying degrees with the Rotate tool, you must type the number when Revit does not display the number you want, the increments shown

   on-screen are set in the _____ dialog box.

9. List all the Snap points available on a circle (ex. Line: endpoint, midpoint,

   nearest) _____.

10. The Snaps Off option must not be checked in the Snaps dialog box to automatically and accurately select snap points while drawing. (T/F)

Notes:

Lesson 4
Drawing 2D Architectural Objects::

This lesson is meant to give you practice drawing in Revit; while doing so you will become familiar with the various shapes and sizes of the more common symbols used in a residential set of drawings.

Don't forget that learning to draw 2D is important in Revit. You often define 3D items, like walls, in 2D floor plans (while Revit takes care of the 3rd dimension automatically). Additionally, you will often want to embellish a view with 2D lines (for example, the exterior elevation; which is a 2D projection of a 3D model).

If you have used other design programs, like AutoCAD or Architectural Desktop, these lessons will help you understand the "different" way to draw using Revit (i.e. a more visual/graphic based input system).

Each drawing will be created in its own *Drafting View*.

For some of the symbols to be drawn, you will have step-by-step instruction and/or a study on a particular tool that would be useful in the creation of that object. Other symbols to be drawn are for practice by way of repetition and do not provide step-by-step instruction.

Do NOT draw the dimensions; they are for reference only.

Exercise 4-1:
Sketching Rectilinear Objects

Overview

All the objects you will draw in this exercise consist entirely of straight lines, either orthogonal or angular.

As previously mentioned, all the objects MUST BE drawn in SEPERATE *Drafting Views*. Each object will have a specific name provided, which is to be used to name the *Drafting View*. All views will be created in one Project file which should be saved in your personal folder created for this course.

view name: **Bookcase**

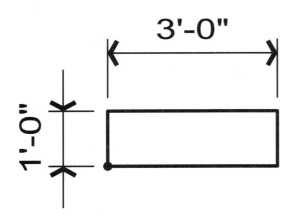

This is a simple rectangle that represents the size of a bookcase.

1. **Open** a new project. **Save** the project as **Ex4-1.rvt**.

2. Create a new *Drafting View* named Bookcase *(from View → New...)*. (Figure 4-1.1)

3. Select **Detail Lines** from the *Drafting* tab.

By default, the current line thickness is set to Fine Lines. You can see this on the *Options Bar* when the Detail Lines tool is active. You will change this to medium next.

4. Set the line thickness to **Medium Lines** (Figure 4-1.2).
   NOTE: Use this setting for all the drawings in this chapter.

5. Click the **Rectangle** button on the *Options Bar*. (Figure 4-1.2)

6. **Draw a rectangle**, per the dimensions shown above, generally in the center of the drawing window.
   TIP: You should be able to accurately draw the rectangle using the temp. dimensions; if not, click on the dimension text to change it.

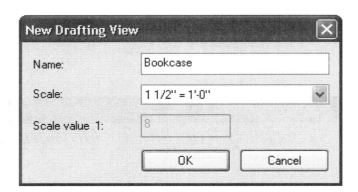

Figure 4-1.1 New Drafting View dialog

Figure 4-1.2 Options Bar: Line thickness set to medium and rectangle button selected

That's it!

view name: **Coffee Table**

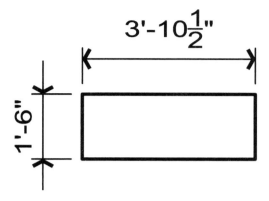

$3'-10\frac{1}{2}"$

1'-6"

7. Create a new *Drafting View*;
    Note: *This step will be assumed for the rest of the drawings in this exercise.*

8. Similar to the steps listed above, plus the suggestions mentioned below, create the coffee table shown to the left.

Draw the rectangle as close as possible using the visual aids; you should be able to get the 1'-6" dimension correct. Then click on the text for the temporary horizontal dimension and type in 3'-10½".

Entering fractions: The 3'-10½" can be entered in one of four ways.

o 3 10.5	*Notice there is a space between the feet and inches*
o 3 10 1/2	*Note the two spaces: they separate the feet, inches and fractions*
o 0 46.5	*This is all in inches; that is, 3'- 10½" = 46.5"*
o 46.5"	*Omit the feet and use the inches symbol.*

view name: **desk-1**

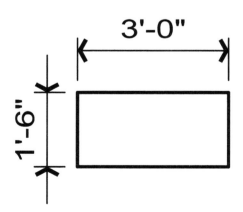

3'-0"

1'-6"

9. Draw the Desk in its own view, similar to the steps outlined above.

TIP: You should double check your drawing's dimensions using the Tape Measure tool. Pick the icon Tools toolbar and then pick two points. A temporary dimension will display.

view name: **Night Table**

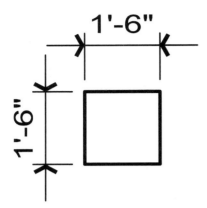

Obviously you could draw this quickly per the previous examples. However, you will take a look at copying a *Drafting View* and then modifying an existing drawing.

You will use the Move tool to stretch the 3'-0" wide desk down to a 1'-6" wide night table.

First you will look at *Drafting Views* and how they are organized in the *Project Browser*, and then you will copy the Desk-1 Drafting View to start the Night Table drawing/view.

10. Locate the **Drafting Views** section in the *Project Browser* and Expand it *(i.e. click on the "plus" symbol)*.

Notice that the three views you previously created are listed (see Figure 4-1.3). The Desk-1 view is shown with bold text because it is the current view. All *Drafting Views* are listed here and can be renamed, copied or deleted at any time by right-clicking on the items in the list. You will copy the Desk-1 view next.

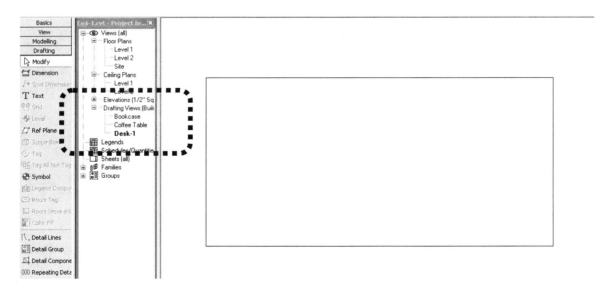

Figure 4-1.3   Project Browser: Drafting Views section expanded

11. Right-click on the Desk-1 view (Figure 4-1.4).

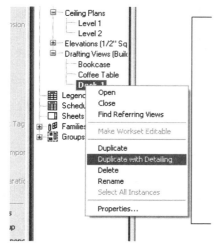

Figure 4-1.4 Drafting Views list–
Right click on Desk-1 view for menu

Next you will select *Duplicate with Detailing* which will copy the *View* and the *View's* contents, whereas the *Duplicate* option will only copy the *View* and its settings but without any line work.

12. Select **Duplicate with Detailing** from the pop-up menu. (Figure 4-1.4)

You now have a new View named *Copy of Desk-1* in the *Drafting Views* list. This *View* is open and current.

13. Right-click on the new view, **Copy of Desk-1,** and select **Rename** from the pop-up menu.

14. Type **Night Table** and then click **OK** (Figure 4-1.5).

Figure 4-1.5 Rename View dialog– Night Table entered

You are now ready to modify the copy of the Desk-1 line work. You will use the Move tool to change the location of one of the vertical lines, which will cause the two horizontal lines to stretch with it.

15. Select the vertical line on the <u>right</u> and then click the **Move** icon on the *Edit* toolbar.

16. Pick the midpoint of the vertical line, move the mouse 1'-6" to the left and then click again. (Figure 4-1.6)

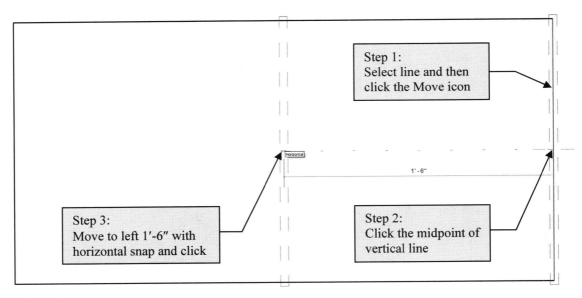

Figure 4-1.6 Move tool used to stretch a rectangle

Figure 4-1.7 Stretch rectangle (using Move tool)

The rectangle is now resized (or better, stretched) to be half its original width.

You will notice that the horizontal lines automatically adjusted.

Revit assumes a special relationship between a line being edited and any line whose endpoint is directly touching it.

17. **Save** your project.

view name: **dresser-1**

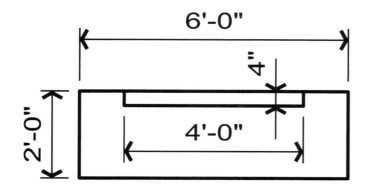

18. Draw the Dresser using the following tip:

    Tip: *Simply draw two rectangles. Move the smaller rectangle into place using the Move tool and Snaps.*

view name: **dresser-2**

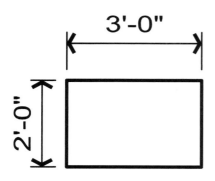

19. Draw this smaller dresser.

view name: **desk-2**

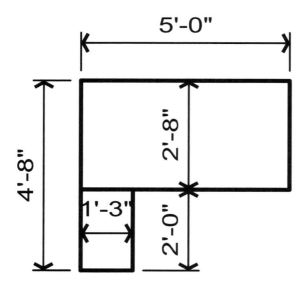

20. Draw this desk in the same way you drew dresser-1 (i.e. two rectangles).

view name: **file cabinet**

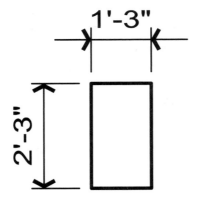

21. Draw this file cabinet

view name: **chair-1**

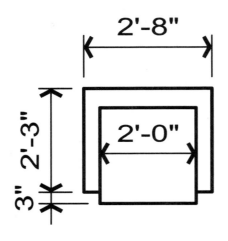

22. Draw this chair per the following tips:

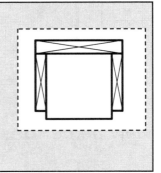

Draw the 2'x2' square first, and then draw three separate rectangles as shown to the right. Move them into place with Move and Snaps. Next, delete the extra lines so your drawing looks like the one shown on the left. Pay close attention to the dimensions!

view name: **sofa**

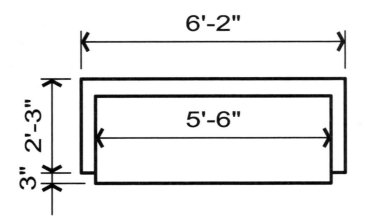

23. *Duplicate* and *Rename* the Chair-1 view to start this drawing.

TIP: See the next page for more info on creating this drawing.

You can create the Sofa in a way similar to how you created the Night Table. That is, you will use the Move tool. Rather than selecting one vertical line, however, you will select all the lines on one side of the chair.

You can select all the lines on one side in a single step (rather than clicking each one individually while holding the Ctrl key). You will select using a Window (not a Crossing Window) selection.

24. **Select** all the lines on the right side (Figure 4-1.8).
   TIP: Pick from left to right for a window selection.

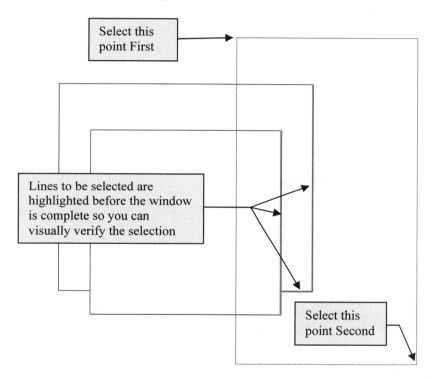

Select this point First

Lines to be selected are highlighted before the window is complete so you can visually verify the selection

Select this point Second

Figure 4-1.8 Stretch chair (using Move tool)

25. Use the **Move** tool to extend the chair into a sofa.
   TIP: The difference between the chair width and the sofa width is the distance the lines are to be moved.

You have just successfully transformed the chair into a sofa! Don't forget to *Save* often to avoid losing work. Also, use the *Tape Measure* tool to verify your drawing's accuracy.

view name: **bed-double.dwg**

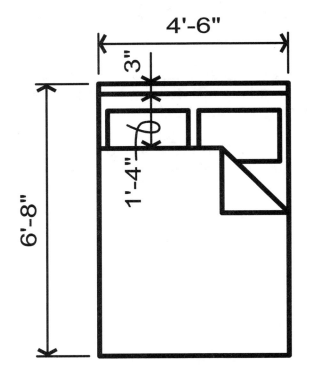

26. Draw this bed using lines and rectangles; the exact size of items not dimensioned is not important.

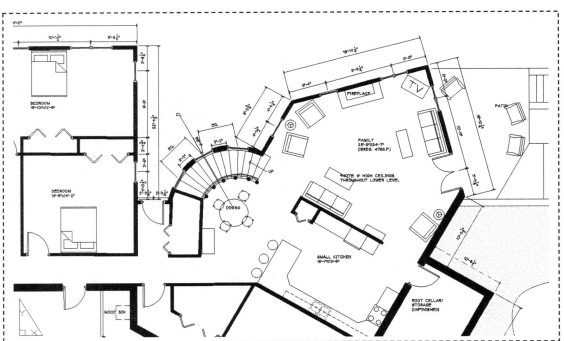

Figure 4-1.9 Floor Plan example using similar furniture to those drawn in this exercise

Image courtesy of Stanius Johnson Architects
www.staniusjohnson.com

Exercise 4-2:
Sketching Objects with Curves

Similar to the previous exercise, you will draw several shapes; this time the shapes will have curves in them.

You will look at the various commands that allow you to create curves, like Arc, Circle and Fillet.

Again, you will not draw the dimensions as they are for reference only; each drawing should be drawn in its own *Drafting View*; name the view with the label provided.

Finally, you can ignore the black dot on each of the drawings below. This dot simply represents the location from which the symbol would typically be inserted or placed.

view name: **laundry-sink.dwg**

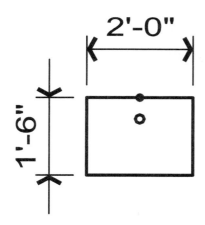

This is a simple rectangle that represents the size of a laundry sink, with a circle that represents the drain.

1. **Open** a new Revit project and name it **Ex4-2.rvt**.

2. Draw the rectangle shown (refer to Exercise 4-1 for more info).

3. Use the **Tape Measure** tool to verify the size of your rectangle.

Next you will draw a circle for the drain. The drain needs to be centered left and right, and 5" from the top edge. The following steps will show you one way in which to do this.

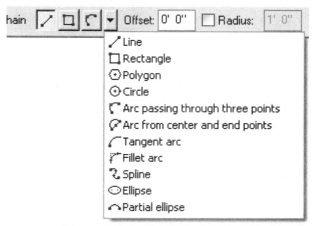

Figure 4-2.1 Sketch Options for detail lines

4. Pick the **Detail Lines** from the *Drafting tab* on the *Design Bar*.

5. Select the **Circle** icon from the *Options Bar* (Figure 4-2.1).
    TIP: *You may need to click the down arrow to see the circle option*

6. Pick the mid-point of the top line (Figure 4-2.2).
    FYI: *This is the center of the circle.*

7. Type **0 1** for the radius and press **Enter** (0 1 = 0′ – 1″).
    FYI: *This creates a 2″ diameter circle.*

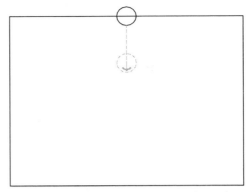

Figure 4-2.2   Creating a circle and moving it into place.

At this point you may get an error indicating the "Element is too small on screen" (Figure 4-2.3). This is supposed to prevent a user, like yourself, from accidentally drawing a very small line when you click the mouse twice right next to each other. If you get this error, you simply zoom in on your drawing more so the circle is more prominently visible on the screen.

Figure 4-2.3   Revit error message when creating the circle IF you are zoomed out too far

8. If you did get the error message just described, click *Cancel*, *Zoom In* and then draw the *Circle* again.

9. Select the circle and then the **Move** tool.

10. **Move** the circle straight down 5″ (Figure 4-2.2)

That's it! Don't forget to save your Ex4-2 project often.

view name: **dryer.dwg**

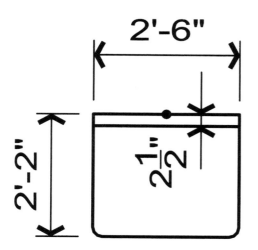

Here you will draw a dryer with rounded front corners. This, like all the other symbols in this lesson, is a plan view symbol (i.e. as viewed from the top).

11. Draw a 30"x26" Rectangle.

12. **Move** the rectangle into place as described in the previous drawing (laundry sink).

Next you will draw the line that is 2½" from the back of the dryer. You will draw the line in approximately the correct position and then use the temporary dimensions to correct the location.

13. Select the **Detail Lines** tool and then pick a point approximately 3" down from the upper-left corner; a temporary dimension will be displayed before clicking your mouse button (Figure 4-2.4).

14. Complete the horizontal line by picking a point perpendicular to the vertical line on the right. (Figure 4-2.4)

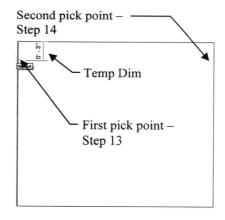

Figure 4-2.4 Drawing a line

15. With the temporary dimensions still displayed, click the text for the 0 - 3" dimension and type **0 2.5** and then press **Enter**. (Figure 4-2.5)

16. Click **Modify** to clear the selection and make the *Temporary Dimensions* go away.

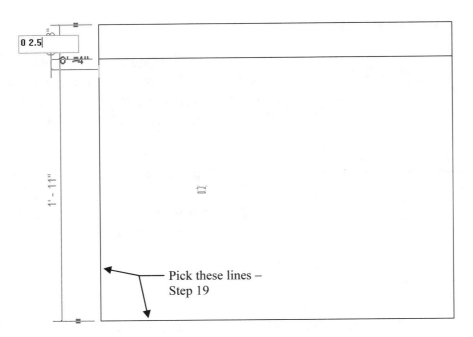

Figure 4-2.5 Adjusting new lines position via temporary dimensions

Next you will round the corners. You still use the *Detail Lines* tool, but use a *Sketch Option* called *Fillet Arc* (pronounced Fill-*it*). You will try this next.

17. With the **Detail Lines** tool active, select the **Fillet Arc** *Sketch Option* (Figure 4-2.1).

18. Select **Radius** and enter **0' 2"** in the text box on the *Options Bar* (Figure 4-2.6).

Figure 4-2.6   Options Bar: options for Detail Lines in Fillet mode

19. Pick the two lines identified in Figure 4-2.5.

The intersection of the two lines is where the arc in placed. Notice the two lines you picked have been trimmed back to the new *Fillet* (arc).

20. Repeat the previous step to **Fillet** the other corner.

view name: **washer.dwg**

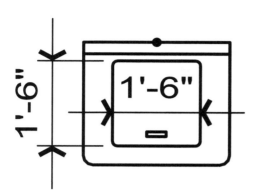

This drawing is identical to the dryer except for the door added to the top. You can duplicate the dryer *Drafting View* to get a jump start on the drawing.

To draw the door you will use the **Offset** option, which is part of the *Detail Lines* tool.

The **Offset** tool is easy to use; to use it you:
- Select the *Detail Lines* tool
- Select the *Pick Lines* icon on the *Options Bar*
- Enter the *Offset* distance on the *Options Bar*
- Select near a line (and on the side to offset)

21. Select the **Detail Lines** tool.

22. Select the **Pick Lines** icon on the *Options Bar* (Figure 4-2.7).

Figure 4-2.7   Options Bar: options for Detail Lines in Pick Lines mode

Note that the *Pick Lines* icon toggles you from *Draw* mode to *Pick Lines* mode. The *Pick Lines* mode allows you to use existing linework to quickly create new linework. You should notice the available options on the *Options Bar* have changed after selecting *Pick Lines*; you can switch back by selecting the Draw icon on the *Options Bar*. (Figure 4-2.7)

23. Enter **0' 6"** for the *Offset* distance (Figure 4-2.7)

Next you will need to select a line to offset. Which side of the line you click on will indicate the direction the line is offset. Revit provides a visual reference line indicating which side the line will be offset to and its location based on the offset distance entered.

Without clicking the mouse button, move your cursor around the drawing and notice the visual reference line. Move your mouse from side-to-side on a line to see the reference line move to correspond to the cursor location.

24. **Offset** the left vertical line towards the right.
    TIP: Make sure you entered the correct offset distance (previous step).
    See Figure 4-2.8.

You will need to look at the washer drawing and the dryer drawing dimensions to figure out the door size (i.e. the offset amount).

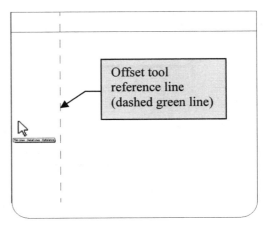

25. Offset the other 3 sides; your sketch should look like Figure 4-2.9.

26. Use the **Fillet Arc** sketch mode, in the *Detail Lines* tool, to create a 1" radius on the four corners.

Figure 4-2.8 Offset left vertical line towards the right 6"

27. Pick anywhere inside the rectangle.

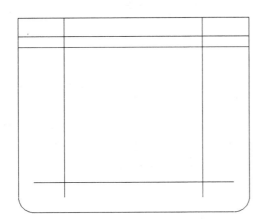

Figure 4-2.9 All four lines offset inward to create outline for washer door

All four corners should now be rounded off.

28. Draw a small 4"x1" rectangle to represent the door handle; draw it anywhere. Using *Snaps*, move it to the midpoint of the bottom door edge, and then move the handle up 2".

29. **Save** your project (make sure the view name is washer).

view name: **range.dwg**

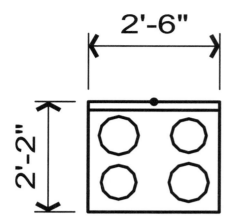

2'-6"

2'-2"

Now you will draw a kitchen range with four circles that represent the burners.

In this exercise you will have to draw temporary lines (called construction lines) to create reference points needed to accurately locate the circles. Once the circles have been drawn the construction lines can be erased.

30. **Draw** the range with a 2″ deep control panel at the back; refer to the steps described to draw the dryer if necessary.

31. **Draw** the four construction lines shown in Figure 4-2.8 (they do not need to be dashed lines).

32. **Draw** two 9½″ diameter circles and two 7½″ diameter circles using the intersection of the construction lines to locate the center of the circles (Figure 4-2.10).
TIP: *Refer back to the laundry-sink for sketching circles.*

7"   1'-4"

11"

7"

Figure 4-2.10 Range with four temporary construction lines (shown dashed for clarity)

33. **Erase** the four construction lines.

34. **Save** your project.

TIP: When using the **Tape Measure** tool, you can select *Chain* on the *Options Bar* to have Revit calculate the total length of several picks. For example, you can quickly get the perimeter of a rectangle by picking each corner. Notice, as you pick points, the *Total Length* is listed on the *Options Bar*.

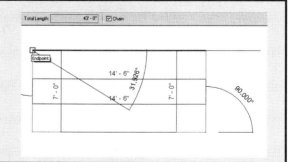

view name: **chair-2.dwg**

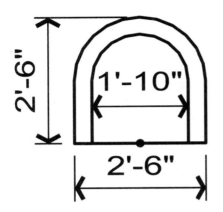

Now you will sketch another chair. You will use *Detail Lines* with *Offset* and the *Arc* modes.

First you will draw the arc at the perimeter.

35. Select *Detail Lines* and then **Arc from center and end points** from the *Options Bar*. (Figure 4-2.1)

Notice on the *Status Bar* you are prompted to "**Click to enter arc center**".

36. *First Pick*: Pick a point somewhere in the middle of a new drafting view named Chair-2 (see Figure 4-2.11).

First point picked (step 36)

Second point picked when temporary dim is 1'-3" (step 37)

You are now prompted on the *Status Bar* to "**Enter arc end point**".

Figure 4-2.11 Sketching an arc; picking 2nd point

37. *Second Pick*: Move the cursor towards the left, while snapped to the horizontal, until the temporary dimensions reads **1'-3"** (see Figure 4-2.11).

NOTE: *As you can see, Revit temporarily displays a full circle (Figure 4-2.11) until you pick the second point. This is because Revit does not know which direction the arc will go or what the arc length will be. The full circle allows you to visualize where your arc will be.*

38. *Third Pick*: Move the cursor to the right until the arc is 180 Degrees and snapped to the horizontal again. (Figure 4-2.12)
    TIP: *Notice that your cursor location determines on which direction/side the arc is created. Move the cursor around before picking the third endpoint to see how the preview arc changes on screen.*

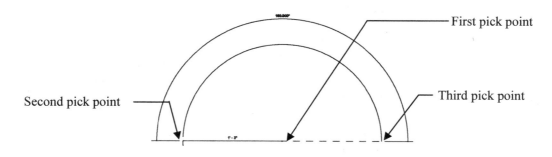

Figure 4-2.12   Pick points for arc command

Next you will draw the three *Detail Lines* to complete the perimeter of the chair.

39. **Draw** two vertical lines **1'-3"** long.
    TIP: Don't forget to use Medium Lines when sketching detail lines. If you forgot, you can select the lines (while in Modify mode) and select Medium Lines from the Type Selector on the Options Bar.

40. **Draw** the **2'-6"** line across the bottom.

Notice the radius is 1'-3". You did not have to enter that number because the three points you picked was enough information for Revit to determine it automatically.

If you had drawn the three lines in steps 39 and 40 before the arc you would not have had the required center point to pick while creating the arc. However, you could have drawn a temporary horizontal line across the top to set up a center pick that would allow you to draw the circle in its final location.

You have now completed the perimeter of the chair.

41. Use the **Offset** tool to offset the arc and two vertical lines the required distance to complete the sketch.

42. **Save** your project.

view name: **Love-Seat.dwg**

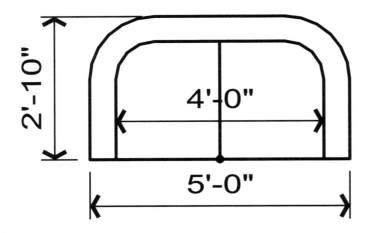

You should be able to draw this Love Seat without any further instruction (use the same radius as Chair-2).

You can do so using a combination of the following tools:
- o Detail Lines
- o Fillet Arc
- o Offset
- o Move

view name: **Tub.dwg**

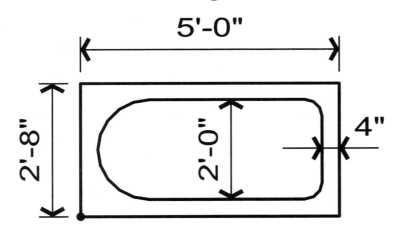

Now you will draw a bath tub. You will use several commands previously covered.

You will use the following tools:
- o Detail Lines
- o Fillet Arc
- o Offset
- o Move

You may also wish to draw one or more construction lines to help locate things like the large radius, although it can be drawn without any.

view name: **Lav-1.dwg**

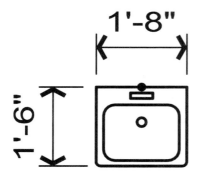

Draw this lavatory per the following specifications:

- o Larger arcs shown shall have a 2½" radius;
- o Smaller arcs shown shall have a 1" radius (outside corners);
- o Sides and Front of sink to have 1½" space (offset);
- o Back to have 4" space (offset);
- o Small rectangle to be 4½" x 1" and 1½" away from the back;
- o 2" diameter Drain is 8" from back.

view name: **Lav-2.dwg**

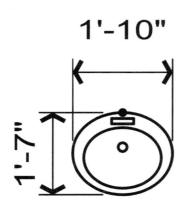

Next you will draw another lavatory. This time you will use the *Ellipse* tool.

43. Create a new Drafting View.

44. Select Detail Lines and then **Ellipse** from the *Options Bar*. (Figure 4-2.1)

Notice the *Status Bar* prompt: "**Click to enter ellipse center**".

45. Pick near the middle of the screen.

46. With **the** cursor snapped to the vertical, point the cursor straight down and then type **0 9.5**, then press **Enter**.
    NOTE: *9½" is half the HEIGHT (1'-7").*

Now you need to specify the horizontal axis of the ellipse.

47. Again, snapping to the horizontal plane, position the cursor towards the right, type **11** and then **Enter**.
    NOTE: *11" is half the width (1'-10").*

That's all it takes to draw an ellipse!

48. **Copy** the ellipse downward (i.e. vertical) **3½"** (Figure 4-2.13).

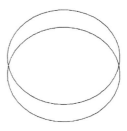

Figure 4-2.13 Ellipse copied downward 3½"

Next you will use the *Resize* tool to scale down the second ellipse. To summarize the steps involved: select the ellipse to be resized, select the resize icon, pick the origin, pick the opposite side of the ellipse, and then enter a new value for the ellipse.

49. Select the second ellipse (i.e. the lower one).

50. Select the **Resize** icon on the *Edit* toolbar.

Notice the *Status Bar* prompt: "**Pick to enter origin**". The origin (or base point) is a point on (or relative to) the object that does not move.

51. First Pick: Pick the top edge (midpoint) of the selected ellipse. (Figure 4-2.14)

52. Second Pick: Pick the bottom edge (midpoint) of the same ellipse. (Figure 4-2.14)

53. Type **1 2** (i.e. 1'-2") and press **Enter**.

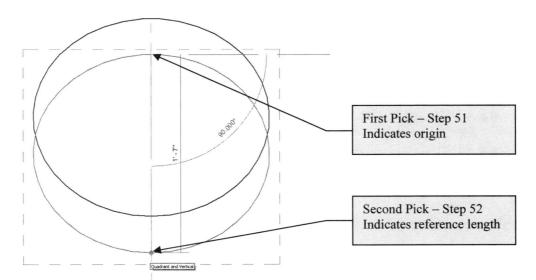

Figure 4-2.14 Resizing ellipse

Next you need to draw the faucet and drain. As it turns out, the faucet and drain for Lav-1 are in the same position relative to the middle/back (black dot). So to save time you will Copy/Paste these items from the Lav-1 drafting view into the Lav-2 view.

54. Open **Lav-1** *Drafting View*.

55. Select the entire Lav-1 sketch.

56. Pick **Copy to Clipboard** from the *Edit* pull-down menu.

57. Switch back to the Lav-2 view and press Ctrl + V on the keyboard (i.e. press both keys at the same time).

58. Pick a point to the side of the Lav-2 sketch (Figure 4-2.15).

59. Select the faucet (i.e. the small rectangle) and the drain (i.e. the circle).

60. Select the **Move** tool; pick the middle/back of the Lav-1 sketch and then pick the middle/back of the Lav-2 sketch.
    NOTE: These two points represent the angle and distance to move the selected items.

61. **Erase** the Lav-1 line work from the Lav-2 view.

You should now have the faucet and drain correctly positioned in your Lav-2.dwg drawing.

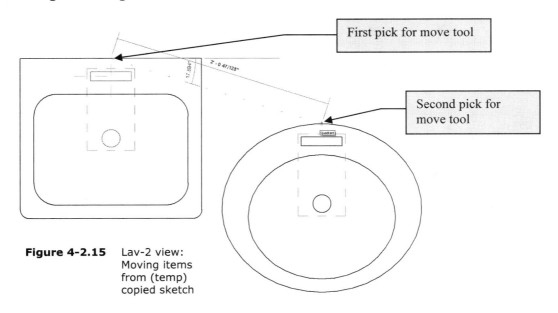

Figure 4-2.15 Lav-2 view: Moving items from (temp) copied sketch

view name: **Sink-1.dwg**

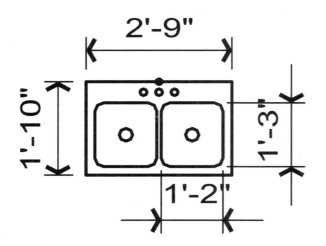

Draw this lavatory per the following specifications:

- o 2" space (i.e. offset) at sides and front
- o 3" diameter Circles centered in sinks
- o 2" radius for Fillets
- o 1½" diameter at faucet; spacing 3½" apart

view name: **Water-closet.dwg**

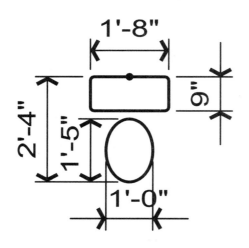

You should be able to draw this symbol without any help.

TIP: *Draw a construction line from the origin (i.e. black dot) straight down 11" (2'-4" – 1'-5" = 11"). This will give you a point to pick when drawing the ellipse.*

view name: **Tree.dwg**

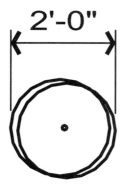

62. Draw one large circle and then copy it similar to this drawing.

63. Draw one small circle (1" diameter) at the approximate center.

64. **Save** your project.

view name: **Door-36.dwg**

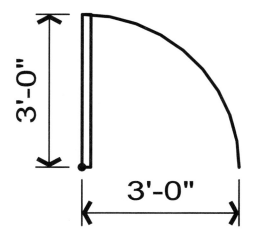

This symbol is used to show a door in a floor plan. The arc represents the path of the door as it opens and shuts.

A symbol like this, 90 degrees open, helps to avoid the door conflicting with something in the house (e.g. cabinets or toilet). Existing doors are typically show open 45 degrees so it is easier to visually discern new from existing.

Revit has an advanced door tool, so you would not actually draw this symbol very often. However, you may decide to draw one in plan that represents a gate in a reception counter or one in elevation to show a floor hatch. Refer back to Figure 4-1.9 for a floor plan example with door symbols.

65. Draw a **2" x 3'-0" rectangle**.

66. Draw an Arc using **Arc from center and endpoints**. Select the three points in the order shown in Figure 4-2.16.
    TIP: *Be sure your third pick shows the cursor snapped to the horizontal plane.*

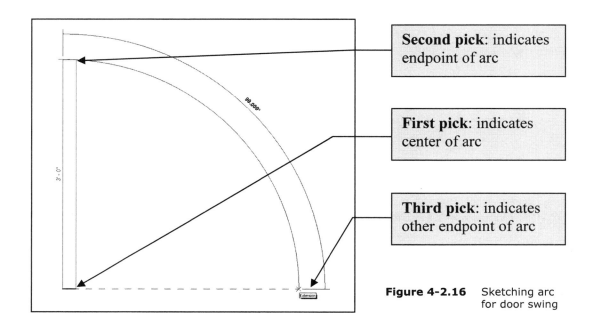

Figure 4-2.16   Sketching arc for door swing

view name: **Door2-36.dwg**

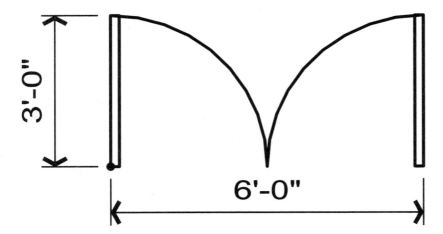

Here you will use a previous drawing and the Mirror tool to quickly create this drawing showing a pair of doors in a single opening.

67. Right-click on **Door-36** view label in the *Project Browser*; select **"Duplicate with detailing"** from the *Context* menu.

68. **Rename** the new *Detail View* **Door2-36**.

Next you will mirror both the rectangle and the arc, but first you will draw a temporary line to use as the *Reflection Line* while in the *Mirror* tool. After mirroring the door you will delete the temporary line.

69. Draw a temporary vertical line that will represent the Line of Reflection (make sure you snap to the endpoint of the arc) needed for the next step. (Figure 4-2.17)

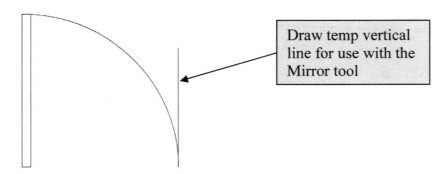

Figure 4-2.17   Temp line required for mirror tool

70. Select the rectangle and the arc (use a Crossing Window by picking from right to left) and then pick the **Mirror** tool from the *Edit Toolbar*.

Notice the *Status Bar* is prompting you to "**Select the axis of reflection**".

71. Move your mouse over the temporary vertical line until it highlights and then click.

At this point the door symbol (i.e. rectangle and arc) is mirrored and the Mirror tool is done. Next you will delete the temporary vertical line as it is no longer needed.

72. Pick the vertical line and press the **Delete** key on the keyboard.

73. **Save** your project (as Ex4-2.rvt).

view name: **clg-fan-1.dwg**

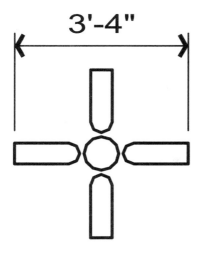

You will use the Array command while drawing this ceiling fan.

The Array command can be used to array entities in a rectangular pattern (e.g. columns on a grid) or in a polar pattern (i.e. in a circular pattern).

You will use the polar array to rotate and copy the fan blade all in one step!

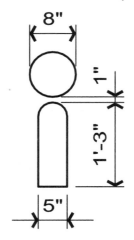

74. Start a new drawing and create the portion of drawing shown in the figure to the right.

75. Select the fan blade (i.e. 1 arc and 3 lines) and then select **Array** from the *Edit* toolbar.

76. Make the following changes to the *Options Bar* (Figure 4-2.18).

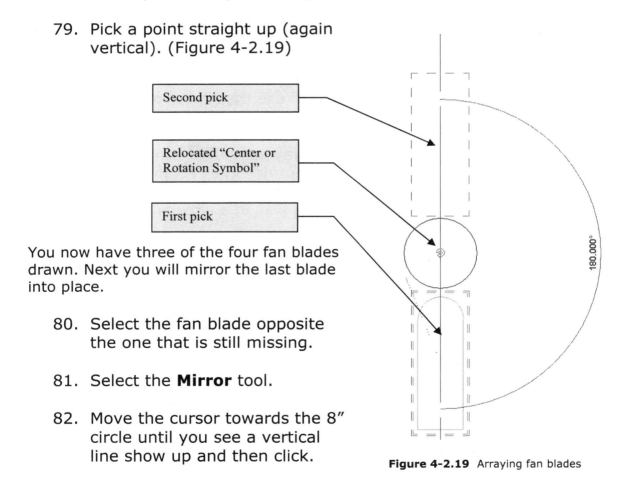

Figure 4-2.18 Array options on the Options Bar

Similar to the Rotate tool, you will relocate the Center or Rotation Symbol to the center of the 8" circle. This will cause the fan blade to array around the circle rather than the center of the first fan blade.

77. *Click* and *Drag* the **Rotation Symbol** to the center of the 8" circle. Make sure Revit snaps to the center point before clicking (see Figure 4-2.19).

78. Pick a point straight down (vertical). (Figure 4-2.19)

79. Pick a point straight up (again vertical). (Figure 4-2.19)

Second pick

Relocated "Center or Rotation Symbol"

First pick

You now have three of the four fan blades drawn. Next you will mirror the last blade into place.

80. Select the fan blade opposite the one that is still missing.

81. Select the **Mirror** tool.

82. Move the cursor towards the 8" circle until you see a vertical line show up and then click.

Figure 4-2.19 Arraying fan blades

The ceiling fan is now complete!

Self-Exam:

The following questions can be used as a way to check your knowledge of this lesson. The answers can be found at the bottom of this page.

1. Revit does not allow you to copy *Detail Views*. (T/F)

2. Entering 4 3.25 in Revit means 4'-3¼". (T/F)

3. Construction lines are useful drawing aids. (T/F)

4. Use the _____ tool to sketch an oval shape.

5. When you want to make a previously drawn rectangle wider you would use the _____ tool.

Review Questions:

The following questions may be assigned by your instructor as a way to assess your knowledge of this section. Your instructor has the answers to the review questions.

1. Use the *Offset* option to quickly create a parallel line(s). (T/F)

2. If Revit displays an error message indicating a line is too small to be drawn, you simply zoom in more and try again. (T/F)

3. Use the _____ command to create a reverse image.

4. With the *Move* tool, lines completely within the crossing-window are actually only moved, not stretched. (T/F)

5. You can relocate the "center of rotation" when using *Radial Array* (T/F)

6. Occasionally you need to draw an object and then move it into place to accurately locate it. (T/F)

7. The _____ (which is an option within the *Details Lines* tool) allows you create a rounded corner where two lines intersect.

8. When using the *Mirror* tool, you occasionally need to draw a temporary line that represents the *Axis of Reflection* (T/F)

9. With line style was required to be selected in the *Type Selector* (on the Options Bar) for all the line work in this Lesson _____.

10. In the Detail Lines tool, how many options allow you to draw arcs? _____

Self-Exam Answers:
1 - F, **2** - T, **3** - T, **4** - Ellipse, **5** - Move

Notes:

Lesson 5
Residence: FLOOR PLAN (First Floor)::

In this lesson you will draw the first floor plan of a single family residence. The project will be further developed in subsequent chapters. It is recommended that you spend adequate time on this lesson as later lessons build on this one.

Exercise 5-1:
Project setup

The Project:

You will model a two-story residence located in a suburban setting. Just to the north of the building site is a medium-sized lake. For the sake of simplicity, the property is virtually flat.

The main entry is from the south side of the building. You enter the building into a large foyer. The second floor has a railing that looks down onto the foyer.

This building is not meant to meet any particular building code. It is strictly a tool to learn how to use Autodesk Revit. Having said that, however, there are several general comments as to how codes may impact a particular part of the design; each must be verified with your local rules and regulations.

Creating the Project file:

A Building Information Model (BIM) consists of a single file. This file can be quite large. For example, the prestigious architectural design firm SOM is using Autodesk Revit to design the Freedom Tower in New York. As you can imagine a skyscraper would be a large BIM file; whereas a single family residence would be much smaller.

Large databases are just starting to enter the architectural design realm. However, banks, hospitals, etc. have been using them for years – even with multiple users!

When Revit is launched, the default template is loaded. This template file has several items set up and ready to use (e.g. some wall, door and window families). Starting with the correct template can save you a significant amount of time.

Revit provides a handful of templates with particular project types in mind. They are Commercial, Construction, Residential and Structural.

In this exercise you will use the Residential template. It has several aspects of the project file pre-setup and ready for use. A few of these items will be discussed momentarily.

As your knowledge in Revit increases you will be able to start refining a custom template (which probably originated from a standard template). The custom template will have things like your firm's title block and a cover sheet with abbreviations, symbols, etc. all set up and ready to go.

Next you will create a new project file.

1. Select **File → New → Project...** from the pull-down menu.

You are now in the *New Project* dialog box. Rather than clicking *OK* you will select *Browse* so you can select a specialized template (Figure 5-1.1).

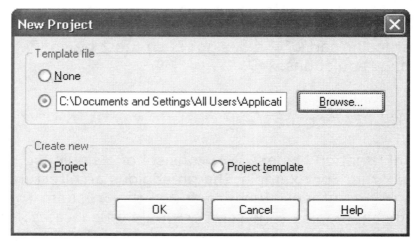

Figure 5-1.1 New Project dialog

2. Click the **Browse...** button.

3.   Select the template named **Residential-Default.rte** from the list of available templates. (Figure 5-1.2)

4.   Click **Open** to select the highlighted template file.

5.   Click **OK** to complete the *New Project* dialog.

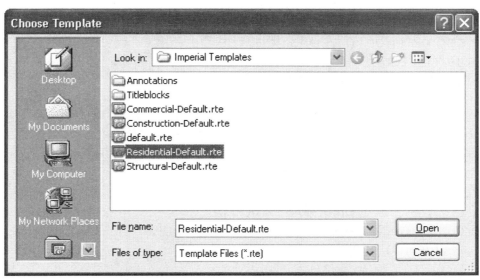

Figure 5-1.2 Choose Template dialog

You have just set up a project file that you will use for the remainder of this book.

Next you will take a look at the predefined wall types that have been loaded.

6.   Select **Wall** from the *Basics* tab on the *Design Bar*.

7.   Click the **Type Selector** down-arrow on the *Options Bar*. (Figure 5-1.3)

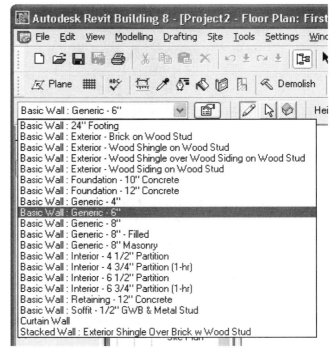

Figure 5-1.3 Preloaded wall types

Wall types:

Notice the wall types that have been preloaded as part of the Residential template. These wall types are many of the types of walls one would expect to find on a typical residential construction project.

The walls are mostly wood stud walls; both interior and exterior. If you look at the Default or Commercial templates you will see mostly metal stud and concrete block (CMU) wall types.

Additionally, the thicknesses of materials can vary between commercial and residential; this is also accounted for in the templates. For example, gypsum board (or sheet rock / wall board / dry wall) is typical 5/8" thick on commercial projects whereas it is typically only 1/2" on residential work.

To reiterate the concept, consider the following comparison between a typical commercial interior wall and a typical residential interior wall.

- <u>Typical Commercial Wall</u>
    - 5/8" Gypsum Board
    - 3 5/8" Metal Stud
    - 5/8" Gypsum Board
        - Total thickness: **4 7/8"**

- <u>Typical Residential Wall</u>
    - 1/2" Gypsum Board
    - 2x4 Wood Stud (3 ½" actual)
    - 1/2" Gypsum Board
        - Total Thickness: **4 1/2"**

Project Browser:

Take a few minutes to look at the *Project Browser* (Figure 5-1.4) and notice the views and sheets that have been set up.

Figure 5-1.4 Project Browser: various items preloaded

Many of the views that you need to get started with the design of a residence are set up and ready to go (Figure 5-1.4).

Practically all the sheets typically found on a residential project have been created; see Figure 5-1.4 under the *Sheets (all)* heading of the project browser. Also, notice the sheets with a "plus" symbol next to them? These are sheets that already have a view (or views) placed on them. (You will study this more later, but a view, such as your East Elevation view, is placed on a sheet at a scale you select.) This means your title block sheets will have printable information as soon as you start sketching walls in one of your plan views.

Project Information:
Revit provides a dialog to enter the basic project information (ex. Client name, project address, etc.). You will enter this information next.

8. From the *Settings* menu select **Project Information**. (Figure 5-1.5)

9. Enter the Project Information shown in Figure 5-1.5.
   FYI: For now you will enter three question marks for the date.

10. Click the **Edit** button next to the *Project Address* parameter. (Figure 5-1.5)

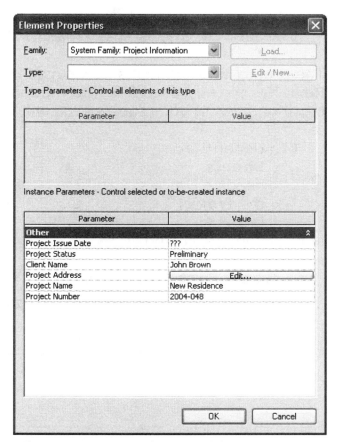

Figure 5-1.5 Project Information dialog

Project Issue Date: ???
Project Status: Preliminary
Client Name: John Brown
Project Name: New Residence
Project Number: 2004-048

11. Enter the project address shown in Figure 5-1.6.
    FYI: This is the address where the house is going to be built, not the client's current address.

Enter:
2138 East Superior Street
Duluth, MN 55812

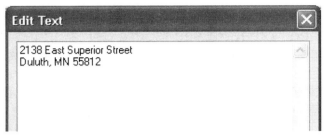

NOTE: *You can enter any address you want to at this point; the address suggested is made up.*

Figure 5-1.6 Adding Project Address (partial view)

12. Click **OK**.

13. Click **OK** to close the *Project Information* dialog box.

The project information is now saved in your Revit project database. Revit has already used this information to infill in a portion of your title block on each of your sheets. You will verify this next.

14. Under the **Sheets** heading, in the *Project Browser*, double-click on the sheet **A1 – First Floor Plan**.

15. Zoom in to the lower right area of title block. (Figure 5-1.7)

Notice that much of the information is automatically filled in.

Your project database is now set up.

16. **Save** your project as **Ex5-1.rvt**.

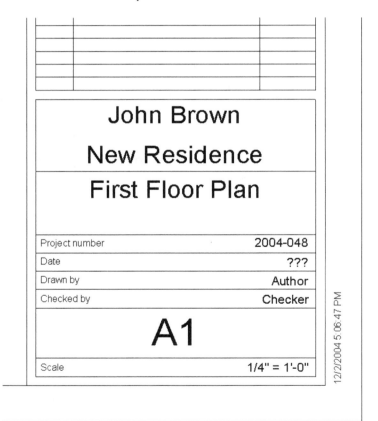

Figure 5-1.7 Sheet A1 – First Floor Plan: Title block with Project Information added automatically

Exercise 5-2:
Exterior Walls

You will begin the first floor plan by drawing the exterior walls. Like many projects, early on you might not be certain what the exterior walls are going to be. So, you will start out using the generic wall styles. Then you will change them to a custom wall style (that you will create) once you have decided what the wall construction is.

Often, a residence is designed with a specific site in mind. However, to keep in line with most drafting and design classes (where floor plans are studied before site plans) you will develop the plans now.

Completed Plans – For Reference:

The following two images show the floor plans as they will look when you are done with this book. You should not need to use any information from these images; they are simply meant to orientate you to the project.

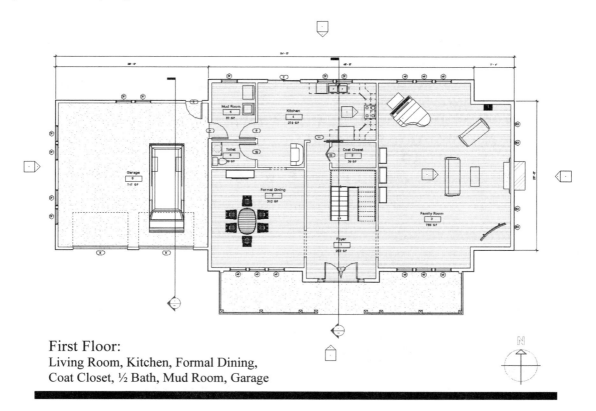

First Floor:
Living Room, Kitchen, Formal Dining,
Coat Closet, ½ Bath, Mud Room, Garage

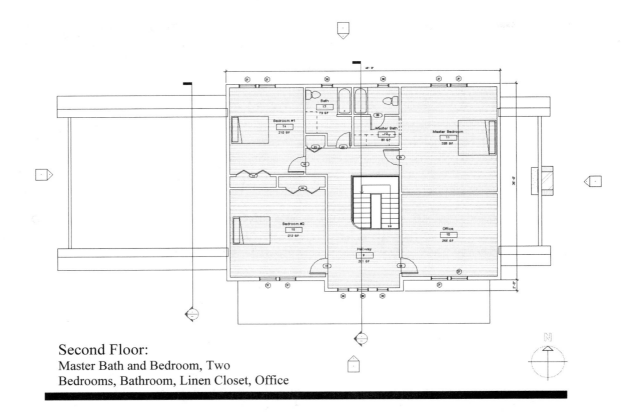

Second Floor:
Master Bath and Bedroom, Two
Bedrooms, Bathroom, Linen Closet, Office

Adjust wall settings:

1. Switch to **First Floor** *View*; select **Wall** form the *Design Bar.*

2. Make the following changes to the wall options in the *Options Bar* (Figure 5-2.1):
    *   Wall style: **Basic Wall: Generic – 8"**
    *   Height: **Unconnected**
    *   Height: **19' 0"**
    *   Loc Line: **Finish Face; Exterior**

Figure 5-2.1 Option bar: Walls

Draw the exterior walls:

3. Draw the walls shown in Figure 5-2.2. Make sure your dimensions are correct. Use the *Tape Measure* tool if you need additional lengths listed. **NOTE**: *If you draw in a clockwise fashion, your walls will have the exterior side of the wall correctly positioned. You can also use the spacebar to toggle which side the exterior face is on.*

TIP: In the *Options Bar*, while you are in the *Wall* tool, you can click *Chain* to continuously draw walls. When *Chain* is not selected you have to pick the same point twice: once where the line ends and again where the next line begins.

The icons to the right of *Chain* allow you to specify what shape wall you wish to draw, just like the 2D sketches in the previous lesson.

Note: *You do not need to draw the dimensions at this point; they are for reference only.*

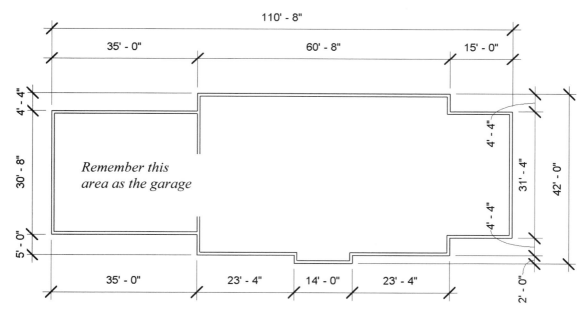

Figure 5-2.2 Exterior walls

Modifying wall dimensions:

Now that you have the exterior walls drawn, you might decide to adjust the building size for masonry coursing (if you have a concrete block foundation), or subtract square footage to reduce cost.

Editing walls in Revit is very easy. You can select a wall and edit the temporary dimensions that appear or you can use the *Move* tool to change the position of a wall. Any walls whose endpoints touch the modified wall are also adjusted (i.e. they grow or shrink) automatically.

Next you will adjust the dimensions of the walls just drawn. You will assume the house will have a concrete block (CMU) foundation.

4. Select the far left wall. (Figure 5-2.3)

5. Select the temporary dimension text (35'-0") and then type **34 8**. (See Figure 5-2.3.) ***FYI***: *Remember you do not need to type the foot or inch symbol; the space distinguishes between them*.

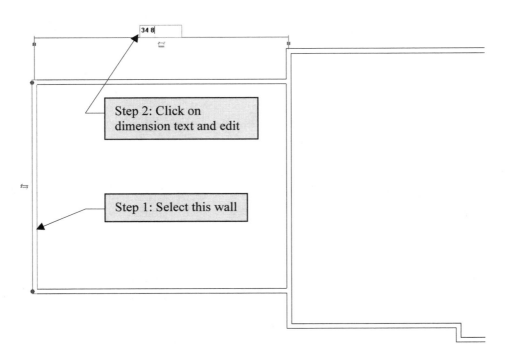

34 8

Step 2: Click on dimension text and edit

Step 1: Select this wall

Figure 5-2.3 Editing wall dimensions

6. In the lower right corner of the floor plan, select the wall shown in Figure 5-2.4.

7. Edit the 4'-4" dimension to 4'-0".
    TIP: *Just type 4 and press enter. Revit will assume feet when a single number is entered.*

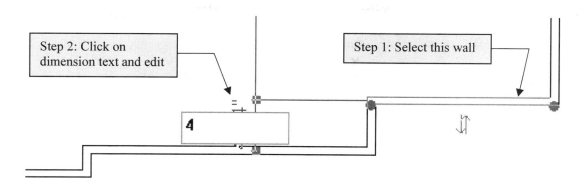

Figure 5-2.4 Editing exterior walls; lower right corner of floor plan

TIP: Concrete blocks (or CMU) come in various widths, and most are 16" long and 8" high. When drawing plans there is a simple rule to keep in mind to make sure you are designing walls to coursing. This applies to wall lengths and openings within CMU walls.

Dimension rules for CMU coursing in floor plans:
- **e'-0"** or **e'-8"**   where **e** is any even number (e.g. 6'-0" or 24'-8")
- **o'-4"**   where **o** is any odd number (e.g. 5'-4")

Using the Align tool with walls:

Revit has a tool called *Align* that allows you to quickly make one wall align with another. You will make the south wall of the garage align with the wall you just modified. After they are aligned you will *Lock* the relationship so the two walls will always move together, which is great when you know you want to walls to remain aligned but might accidentally move one (maybe while zoomed in and you cannot see the other wall).

8. From the *Tools* toolbar, select the **Align** icon.

Notice the *Status Bar* is asking you to select a reference line or point. This is the wall (or linework) that is in the correct location; the other wall (walls) will be adjusted to match the reference plane.

9. Select the exterior face of the horizontal wall you just modified. (Figure 5-2.5)

You should notice a temporary dashed line appear on screen. This will help you to visualize the reference plane.

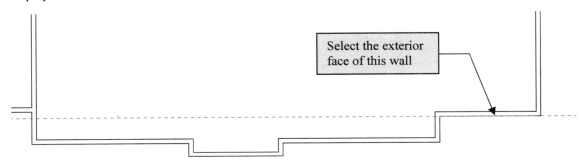

Select the exterior
face of this wall

Figure 5-2.5 Align tool: wall selected

Notice the *Status Bar* is prompting you to select entities to align with the temporary reference plane.

10. Now, select the exterior face of the south (bottom) garage wall. (Figure 5-2.6)

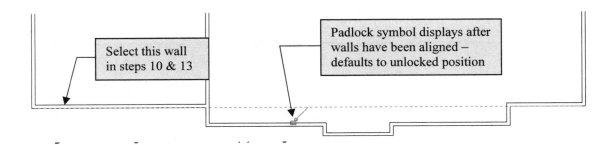

Select this wall
in steps 10 & 13

Padlock symbol displays after
walls have been aligned –
defaults to unlocked position

Note that if you would have selected the interior face or the wall, that side of the wall would have aligned with the reference plane rather than the exterior as it is.

Also, notice that you have a padlock symbol showing. Clicking the padlock symbol will cause the alignment relationship to be maintained. Next you will lock the alignment relationship and experiment with a few edits to see how modifying one wall affects the other.

11. Click on the **padlock** symbol. (Figure 5-2.6)

The padlock symbol is now in the locked position.

12. Click the **Modify** tool to deselect the walls.

13. Select the garage wall identified in Figure 5-2.6.

Next you will make a dramatic change so you can clearly see the results of that change on both "locked" walls.

14. Change the 4'-0" dimension to **20'-0"**. (Figure 5-2.7)

Notice that both walls moved together (Figure 5-2.7). Also notice that when either wall is selected the padlock is displayed, which helps in identifying "locked" relationships while you are editing the project. Whenever the padlock symbol is visible, you can click on it to unlock it (or remove the aligned relationship).

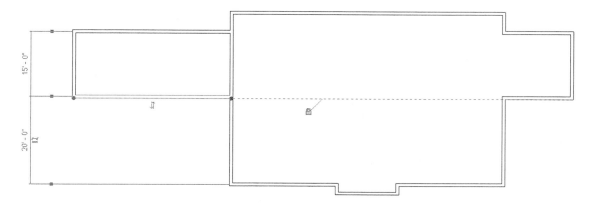

Figure 5-2.7 Wall edit: Both walls move together

Next you will make a few more wall modifications, on your own, to get the perimeter of the building to masonry coursing.

15. Modify wall locations to match Figure 5-2.8; do not draw the dimensions at this time.

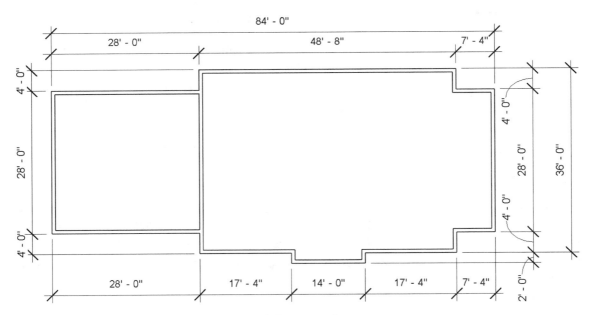

Figure 5-2.8 Exterior wall dimensions (final dimensions)

Changing a wall type:

In the next several steps, you will learn how to change the generic walls into a more refined wall type. Basically, you select the wall (or walls) you want to change and then you pick a different wall type from the type selector on the *Options Bar*. Remember, the *Options Bar* is context sensitive; which means the wall types are only listed in the *Type Selector* when you are using the *Wall* tool or you have a wall selected.

16. Select all the walls in your First Floor Plan view.
   TIP*: You can select all the walls at once. Click and drag the cursor to select a window. Later in the book you will learn about Filters which also aid in the selection process.*

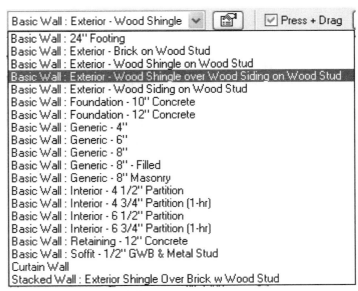

Figure 5-2.9 Type Selector

17. From the *Type Selector* on the *Options Bar*, select **Basic Wall: Exterior – Wood Shingle over Wood Siding on Wood Stud** (Figure 5-2.9).

18. Click **Modify** to deselect the walls.

If you zoom in on the walls you can see their graphic representation has changed a little. It would have changed a lot, of course, if you had selected a brick and CMU cavity wall. But you typically try to start with a wall closest to the one you think you will end up with (you may never even have to change the wall type).

Next you will create a custom wall type for the walls around the garage. You will need a wall type similar to the one you just selected, but with 2x4 studs rather than 2x6's. You will start by making a copy of the existing wall type and then modify the copy.

Create a custom wall style:

As previously mentioned, Revit provides several predefined wall styles, from metal studs with gypsum board to concrete block and brick cavity walls. However, you will occasionally need a wall style that has not yet been predefined by Revit. You will study this feature next.

First, you will take a close look at the predefined wall type you are using for the exterior walls.

19. With the *Wall* tool selected, pick the wall type: **Basic Wall: Exterior – Wood Shingle over Wood Siding on Wood Stud** from the *Type Selector* drop-down list (Figure 5-2.9).

20. Click the **Properties** button to the right of the *Type Selector*.

21. You are now in the *Element Properties* dialog box. Click the **Edit / New** button. (Figure 5-2.10)

22. You should be in the *Type Properties* dialog box. Click the **Edit** button next to the *Structure* parameter. (Figure 5-2.11)

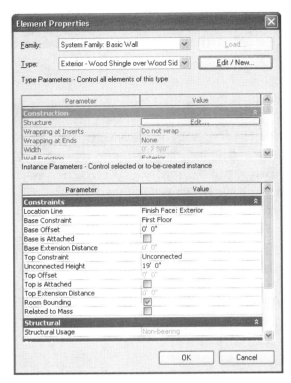

Figure 5-2.10 Element Properties

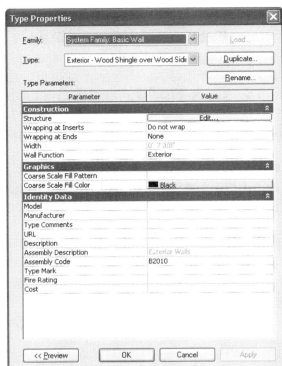

Figure 5-2.11 Type Properties

23. Finally, you are in the Edit Assembly dialog box. This is where you can modify existing wall types or create new ones. Click **<<Preview** to display a preview of the selected wall type. (Figure 5-2.12)

Here, the *Edit Assembly* dialog box allows you to change the composition of an existing wall or to create a new wall.

Things to notice (Figure 5-2.12):
- The exterior side is labeled at the top and interior side at the bottom.
- You will see horizontal lines identifying the core material. The core material can be used to place walls and dimension walls. For example, the wall tool will let you draw a wall with the interior or exterior core face as the reference line. On an interior wall you might dimension to the face of the stud rather than to the finished face of gypsum board; this would give the contractor the information needed for the part of the wall he will build first.
- Each row is called a layer. By clicking on a layer and picking the **Up** or **Down** buttons, you can reposition materials within the wall assembly.

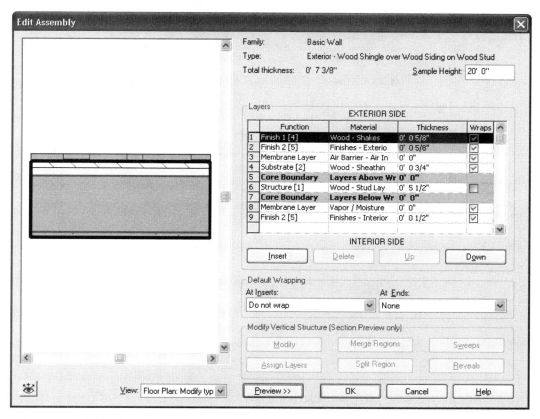

Figure 5-2.12 Edit Assembly

24. Click **Cancel** in each open dialog box to close it.

25. Make sure the wall type: **Basic Wall: Exterior – Wood Shingle over Wood Siding on Wood Stud** is still selected in the type selector.

26. Click the **Properties** button next to the type selector.

27. Click the **Edit / New** button.

28. Click **Duplicate**.

29. Enter **Exterior – Wood Shingle over Wood Siding on 2x4 Wood Stud** for the new wall type name, and then click **OK**. (Figure 5-2.13)
   FYI: Notice that you have simply added "2x4" as a descriptor, making the name distinguishable from the others.

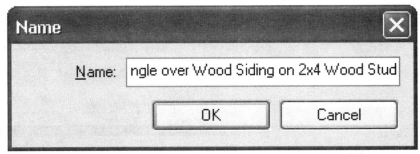

Figure 5-2.13 New wall type name

30. Click the **Edit** button next to the *Structure* parameter.

Next you will simply change the size of the wood stud layer and then save your changes.

31. Change the *Thickness* of the *Wood – Stud Layer* to **0' 3½"**.

32. Your dialog should look like **Figure 5-2.14**. Click **OK** to close all dialog boxes.

You now have access to your newly defined wall. You can sketch new walls using the Wall tool or you can change existing walls to your new wall type using the Modify tool.

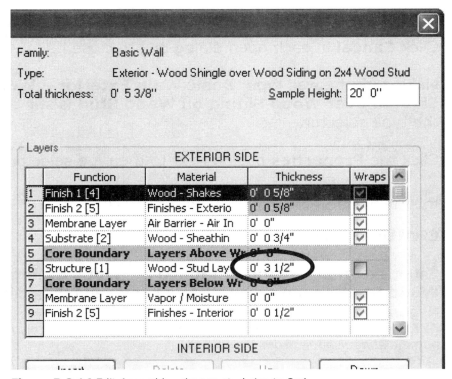

Figure 5-2.14 Edit Assembly; change stud size to 2x4

The next step is to change the wall type for the three walls around the garage.

33. Select the **Modify** button from the *Design Bar*; this allows you to select objects in your drawing.

34. **Zoom out** so you can see the entire plan. Dragging your mouse from upper-right to lower-left, make a crossing-window to select the three garage walls.

35. With the walls selected, pick **Basic Wall: Exterior – Wood Shingle over Wood Siding on 2x4 Wood Stud** from the *Type Selector* drop-down (see tip below).

You should notice the wall thickness change; the reference line (face, center, etc.) does not move. You had Finish Face: Exterior selected when you drew the walls; so the exterior side of the wall does not move.

> **TIP:** If, after selecting all the walls, the *Type Selector* is not active and does not show any wall types, you probably have some other object selected like text or dimensions (although you have not been instructed to draw anything but walls yet). Try to find those objects and delete them.
>
> You can also click on the *Filter* button (located on the *Options Bar* when objects are selected) and uncheck the types of objects to exclude from the current selection.

For many of the walls, Revit can display more refined linework and hatching within the wall. This is controlled by the *Detail Level* option for each view. At the moment, making this change will only reveal the exterior sheathing, but this is helpful information. You will change the *Detail Level* for the *First Floor Plan* View next.

36. Click on the **Detail Level** icon in the lower-left corner of the Drawing Window (Figure 5-2.15).

Figure 5-2.15 Detail Level; Set to medium

37. Select **Medium**.

You may need to zoom in, but you should now see a line added for the exterior sheathing. If you did not pay close enough attention when drawing the walls originally, some of your walls may show the sheathing to the inside of the building.

38. Select **Modify** (or press **Esc**); select a wall. You will see a symbol appear that allows you to flip the wall orientation by clicking on that symbol. (Figure 5-2.16)

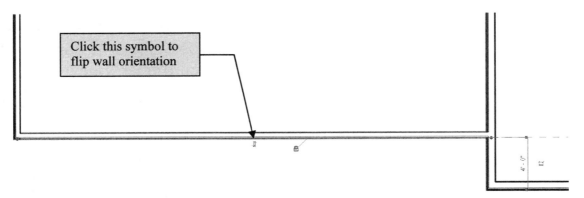

Figure 5-2.16 Selected Wall

39. Whether you need to adjust walls or not, click on the flip symbol to experiment with its operation.
    TIP: The flip symbol is always on the exterior side (or what Revit thinks is the exterior side) of the wall.

40. If some walls do need to be adjusted so the sheathing is to the exterior, do it now. You may have to select the wall(s) and use the *Move* tool to reposition the walls to match the required dimensions.

TIP: You can use the MOVE tool on the menu bar to accurately move walls.

Follow these steps to move an object:
- Select the wall
- Click the Move icon
- Pick any point on the wall
- Start the mouse in the correct direction (don't click)
- Start typing the distance you want to move the wall and press Enter.

You can see your progress nicely with a 3D view. Click the **Default 3D View** button. The walls are shaded on the exterior side to make the image read better. (Figure 5-2.17)

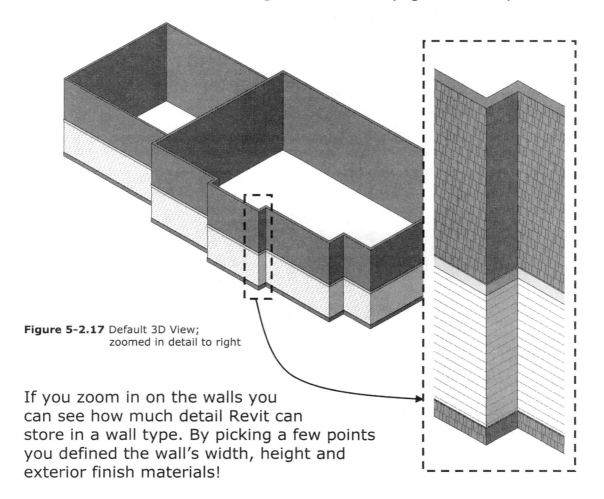

Figure 5-2.17 Default 3D View; zoomed in detail to right

If you zoom in on the walls you can see how much detail Revit can store in a wall type. By picking a few points you defined the wall's width, height and exterior finish materials!

The last thing you will do in this exercise is adjust the height of the walls at the garage and a portion of the living room (refer to image on book cover). Revit allows you to modify the building model in any view. You will make the wall height adjustments in the current 3D View (per the following steps).

41. In the 3D View, click to select the three exterior garage walls.
    TIP: Holding down the Ctrl key allows you to select multiple objects, while holding down the Shift key allows for subtraction from the current selection set.

You will notice, similar to other views, when an object is selected in a 3D view the entire object turns red. This helps to see which objects are in the current selection set when several items are selected.

You should also see the *Options Bar* has been populated with options for the selected walls, just like the plan views. You will adjust the height setting in the *Properties Dialog* next.

42. Click on the **Properties** button (next to the type selector) on the *Options Bar*.

43. Change the *Unconnected Height* parameter to **12'-0"** and then click **OK** to close the dialog.

44. Make the same height adjustment to the three exterior walls at the easternmost portion of the living room (Figure 5-2.18).

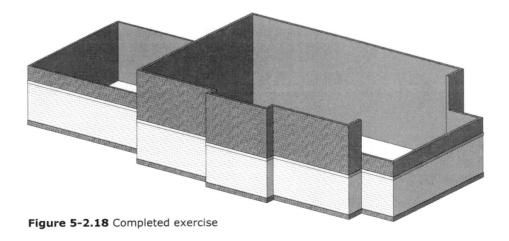

Figure 5-2.18 Completed exercise

Save your project as **ex5-2.rvt**.

Exercise 5-3:
Interior walls

In this lesson you will draw the interior walls for the first floor. Using the line sketching and editing techniques you have studied in previous lessons, you should be able to draw the interior walls with minimal information.

Overview on how plans are typically dimensioned:

The following is an overview of how walls are typically dimensioned in a floor plan. This information is intended to help you understand the dimensions you will see in the exercises, as well as preparing you for the point when you dimension your plans (in a later lesson).

Stud walls (wood or metal) are typically dimensioned to the center of the walls. This is one of the reasons the walls do not need to be the exact thickness. Here are a few reasons why you should dimension to the center of the stud rather than to the face of the gypsum board:

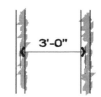

Dimension Example
- Two Stud Walls

- o When the contractor is laying out the walls in a large "empty" area, the most useful dimension is to the center of the stud; that is where they will make a mark on the floor. If the dimension was to the face of the gypsum board, the contractor would have to stop and calculate the center of the stud, which is not always the center of the wall thickness (e.g. a stud wall with 1 layer of gypsum board on one side and 2 layers of gypsum board over resilient channels on the other side of the stud).
- o When creating a continuous string of dimensions, the extra dimensions (text and arrows) indicating the thickness of the walls would take up an excessive amount of room on the floor plans; space that would be better used by notes.

Occasionally you should dimension to the face of a wall rather than the center; here's one example:
- o When indicating design intent or building code requirements, you should reference the exact points/surfaces. For example, if you have uncommonly large trim around a door opening, you may want a dimension from the edge of the door to the face of the adjacent wall. Another example would be the width of a hallway; if you want a particular width you would dimension between the

two faces of the wall and add the text "clear" below the dimension to make it known, without question, that the dimension is not to the center of the wall.

Dimensions for masonry and foundation walls:
- Foundation and masonry walls are dimensioned to the nominal face and not the center. These types of walls are modular (e.g. 8"x8"x16") so it is helpful, for both designer and builder, to have dimensions that relate to the masonry wall's "coursing".

Dimension Example
- Two masonry Walls

Dimensions from a stud wall to a masonry wall:
The rules above apply for each side of the dimension line. For example, a dimension for the exterior foundation wall to an interior stud wall would be from the exterior face of the foundation wall to the center of the stud on the interior wall.

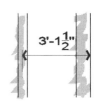

Dimension Example
– stud to masonry

Again, you will not be dimensioning your drawings right away, but Revit makes it easy to comply with the conventions above.

Adjust wall settings:

1. Select **Wall** form the *Design Bar*.

2. Make the following changes to the wall options in the *Options Bar* (Figure 5-2.1):
   - Wall style: **Basic Wall: Interior – 4½" partition**
   - Height: **Second Floor**
   - Loc Line: **Wall Centerline**

Contos Residence

Image courtesy of Anderson Architects
Alan H. Anderson, Architect, Duluth, MN

Drawing the interior walls:

Now that you have the perimeter drawn, you will draw a few interior walls. When drawing floor plans it is good practice to sketch out the various wall systems you think you will be using; this will help you determine what thickness to draw the walls. Drawing the walls at the correct thickness helps later when you are fine-tuning things.

Although Revit provides most of the wall types required, it is a good idea for the student to understand exactly what is being drawn.

Take a moment to think about the basic interior wall in a residential project; it typically consists of the following:
- 2x4 wood studs
  (actual size is 1½" x 3½")
- ½" gypsum board

The wall system described above is sketched to the right (Figure 5-3.1).

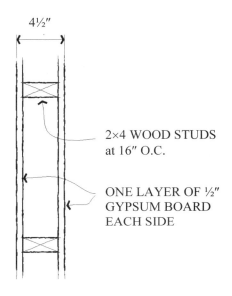

4½"

2×4 WOOD STUDS
at 16" O.C.

ONE LAYER OF ½"
GYPSUM BOARD
EACH SIDE

Figure 5-3.1
Sketch – Plan view of interior wall system

All your interior walls should be drawn **4½"** <u>for each floor</u>.

WOOD STUD WALL

Image courtesy of
Stanius Johnson Architects
Daniel's Construction, Amery WI.
Photo by Robert Aho
www.staniusjohnson.com

3. Draw a vertical wall approximately as shown in Figure 5-3.2. You will adjust its exact position in the next step (step #4).

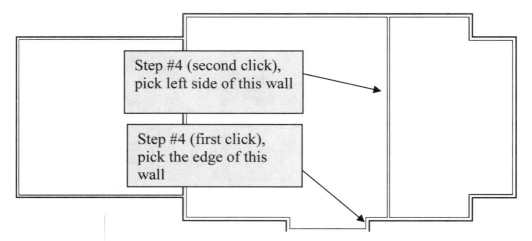

Step #4 (second click), pick left side of this wall

Step #4 (first click), pick the edge of this wall

Figure 5-3.2 First interior wall

4. Use the **Align** tool to align the interior wall you just drew with the edge of the exterior wall (Figure 5-3.2). When you are done, the wall should look like Figure 5-3.3.

5.

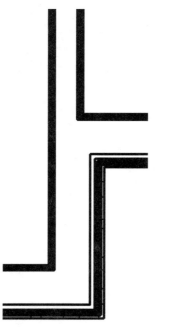

6. Create the same wall for the west (left) side of the entry, repeating the above steps.

Next you will draw a handful of walls per the given dimensions. You will use the same partition type (4½" partition).

7. Draw the interior walls shown in Figure 5-3.4. See the tips below which should help you sketch the walls more accurately.

Figure 5-3.3 Aligned wall

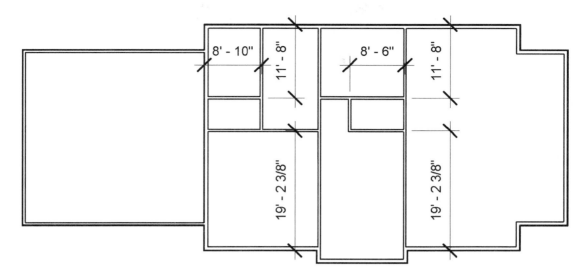

Figure 5-3.4 Layout dimensions for first floor interior walls

TIP – **Dimension witness line control:**

When sketching walls, you can adjust the exact position of the wall after its initial placement. However, the temporary dimensions do not always reference the desired wall location (left face, right face, and center). You will see how easy it is to adjust a dimension's witness line location so you can place the wall exactly where you want it.

First, you will be introduced to some dimension terminology. The two "boxed" notes below are only for permanent dimensions, the others are for both permanent and temporary.

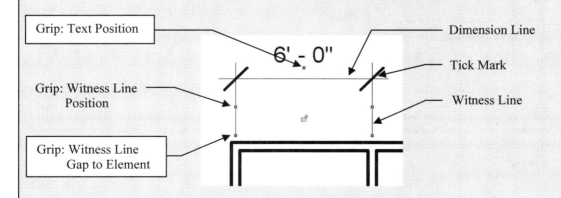

Next you will adjust the witness lines so you can modify the inside clear dimension (between the walls) in the above illustration.

continued on next page

TIP – Dimension witness line control...continued from previous page

When you select a wall you will see temporary dimensions, similar to the image below. **NOTE**: *Some elements do not display temporary dimensions; in that case you can usually click the "Activate Dimensions" button on the Options Bar to see them.*

Clicking on the witness line grip (see image below) causes the witness line to toggle between the location options (left face, right face, center). In the example below, the grip for each witness line has been clicked until both witness lines refer to the "inside" of the room.

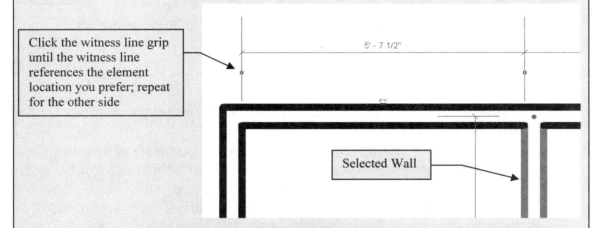

Click the witness line grip until the witness line references the element location you prefer; repeat for the other side

Selected Wall

In the above example the temporary dimension could now be modified, which would adjust the location of the wall to a specific inside or clear dimension.

Not only can you adjust the witness line about a given element, you can also move the witness line to another element. This is useful because Revit's temporary dimensions for a selected element do not always refer to the desired element. You can relocate a witness line (temporary or permanent) by clicking and dragging the grip.

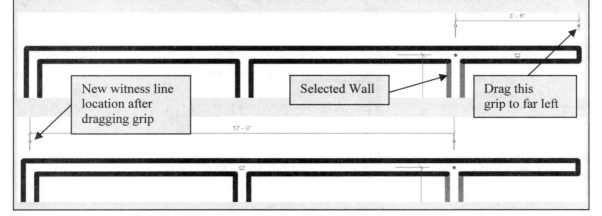

New witness line location after dragging grip

Selected Wall

Drag this grip to far left

Next you will use the *Trim* tool to remove a portion of wall that is not needed.

8.   Select **Trim** from the *Tools* toolbar.

9.   Select the two lines identified in Figure 5-3.5.

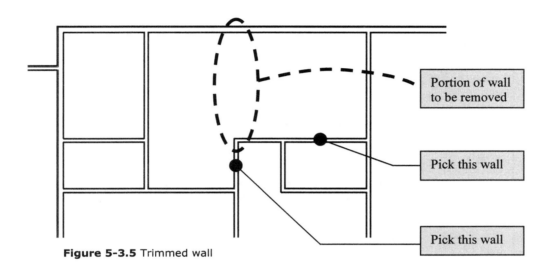

Portion of wall
to be removed

Pick this wall

Pick this wall

Figure 5-3.5 Trimmed wall

Modify an existing wall:

Next you want to change the portion of wall between the house and the garage. Currently, the wall is your typical exterior wall which includes siding on the garage side. The entire vertical wall is one element, so to only change the interior portion you will need to split the current wall, trim the corners and then draw a new interior wall.

10. **Zoom** in on the interior garage wall and select the **Split** tool.

11. Pick somewhere in the middle of the wall. (Figure 5-3.6a)

12. Use the **Trim** tool to trim the corners so the exterior wall only occurs at exterior conditions. (Figure 5-3.6b)
    TIP: *Select the portion of wall you wish to retain.*

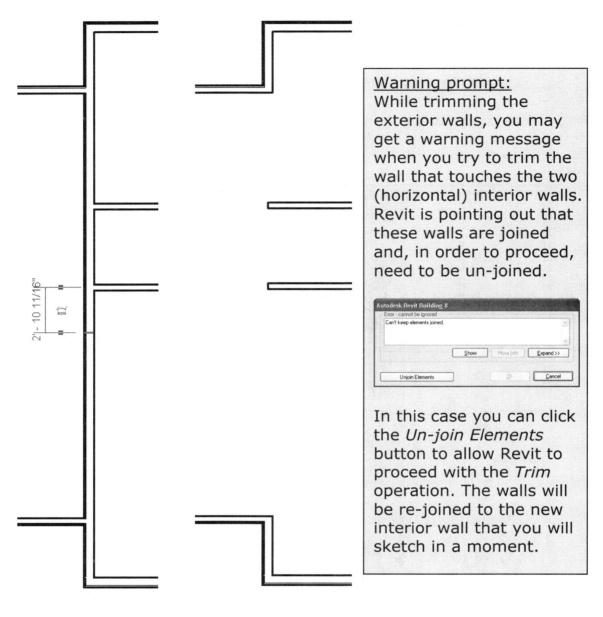

Warning prompt:
While trimming the exterior walls, you may get a warning message when you try to trim the wall that touches the two (horizontal) interior walls. Revit is pointing out that these walls are joined and, in order to proceed, need to be un-joined.

In this case you can click the *Un-join Elements* button to allow Revit to proceed with the *Trim* operation. The walls will be re-joined to the new interior wall that you will sketch in a moment.

Figure 5-3.6a Split wall   **Figure 5-3.6b** Trim wall

Additional custom wall types:

Next you will create a new interior partition type using a copy of the original exterior wall type as a starting point. It is easier to start with a wall that is similar to the one you wish to create.

Structurally, this wall could probably get by with 2x4 studs. However, directly above this wall will be an exterior wall (the portion above the garage roof) that needs to have 2x6 studs to allow for adequate insulation. Additionally, you will want to insulate the interior garage wall; this helps with noise and heat when the large garage door is opened in -20 degree weather!

Truthfully, in this case it may be easier to start with a new wall, given the complexity of the exterior wall you are about to modify. However, this will give you a little more insight on how the exterior wall is constructed and the inner workings of Revit's wall assembly.

13. Using wall type: **Basic Wall: Exterior – Wood Shingle over Wood Siding on Wood Stud** as a starting point, create a new wall type named **Exterior – Garage Wall (Interior)**. *(Remember to click Properties, Edit/New and then Duplicate.)*

Typically you will want to give walls more generic names so they can be used elsewhere in the project. Because you know this wall will only be used once it is easier to give it a more descriptive name so it is not confused with other wall types later.

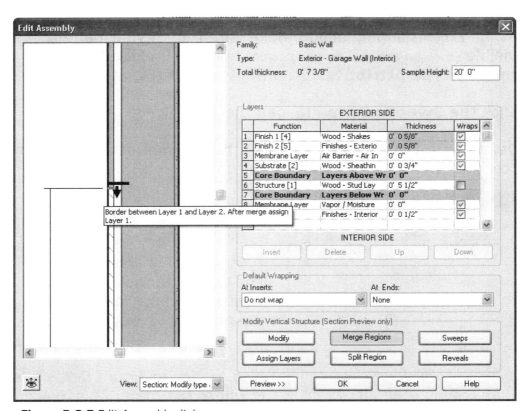

Figure 5-3.7 Edit Assembly dialog

The Edit Assembly dialog shown in Figure 5-3.7 will be modified in the following way. (This is just an overview; the detailed instructions will follow.)

- First, you will delete the horizontal sweeps.
  - A sweep is a closed 2D shape that is extruded along the length of a wall at a given height.
  - This wall type currently has two sweeps (refer back to Figure 5-2.17): one at the base of the wall and another near the second floor line.

- Second, you will merge the two "vertically stacked" exterior finish layers so you can delete the exterior finish layers.
  - A layer can be split vertically into multiple regions; the exterior layer of this wall style is split into two regions. One has the material "Wood-Shakes" assigned to it and the other has "Finishes – Exterior – Siding / Clapboard" assigned to it.
  - Looking at Figure 5-3.7, you can visually see which layers are spilt vertically by looking at the Thickness column; the shaded "Thickness" cells are "vertically stacked" layers.
  - Vertically stacked layers must be merged back into one region before they can be deleted from the layers list.

- Finally, you will change the plywood sheathing to gypsum board and move the air barrier between the gypsum board and the studs,

14. In the *Edit Assembly* dialog make sure the *Preview* pane is visible (*click on the <<Preview button*) and then select **Section: modify type attributes** from the *View* list (Figure 5-3.7).

15. Click the **Sweeps** button to load the *Wall Sweeps* dialog.

16. Delete both *Wall Sweeps.* Click on the row number to select the row and then click the **Delete** button (Figure 5-3.8).

The two rectangular shaped sweeps are now removed from the preview section.

Also, notice in Figure 5-3.8 this is where the material is assigned to the sweep.

Figure 5-3.8 Wall Sweeps to be deleted

Next you will merge the two "vertically stacked" layers into one region. The tricky part here is that Revit will let you merge any two regions. So you need to keep an eye on the cursor and the pop-up tooltip.

17. Select the **Merge Regions** button.

18. Using the *scroll wheel* on the mouse, zoom in on the region split line (at the top edge of the vertical dimension).

19. Move your cursor over the short horizontal line and then press the **Tab** key until the cursor changes to a vertical line with an arrow pointing downward and the tooltip states that Layer 1 and Layer 2 will be merged (see Figure 5-3.7).

20. Individually select the *Layer* numbers 1 and 2 (to select the row) and **Delete** them (this will delete the two exterior finish layers: shakes and siding).

21. Click the **Membrane Layer** row and then click the **Down** button to move the location of the air barrier within the wall.

22. Change the ¾" sheathing layer by clicking in each cell in that row and then clicking the down-arrow that appears to the right. Change the *Function* to Finish 1 [4], *Material* to Finishes – Interior – Gypsum Wall Board, *Thickness* to ½" (Figure 5-3.9).

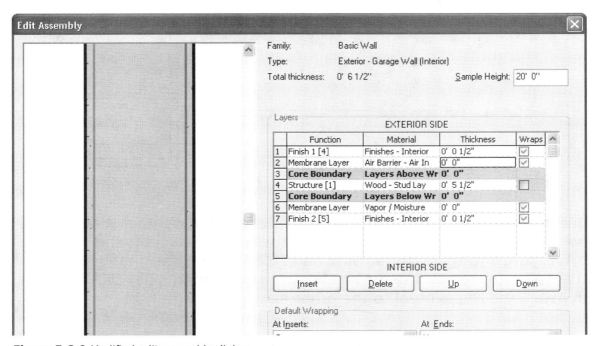

Figure 5-3.9 Modified edit assembly dialog

Now that you have created a new wall type, you can sketch the wall in the first floor plan view.

23. With your new wall type selected in the *Type Selector*, use the **Wall** tool to sketch a wall between the garage and the house; set the wall *Height* to **Second Floor**.
    TIP: Set the Location Line to Finish Face: Interior and then select the interior corner of the exterior wall.

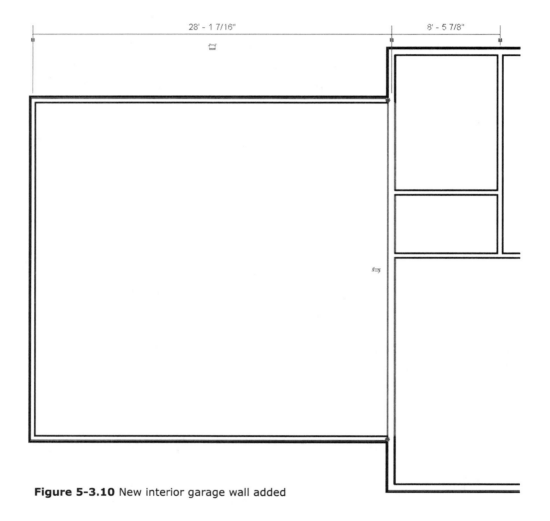

Figure 5-3.10 New interior garage wall added

It is important to remember that the "flip control" symbol is displayed on the exterior side of the wall when the wall is selected; this corresponds to the Exterior Side and Interior Side designations with the wall type editor. If your symbol is on the wrong side, click it to flip the wall and the symbol to the other side.

24. Save your project as **Ex5-3.rvt**.

Exercise 5-4:
Doors, Openings and Windows

This lesson will take a closer look at inserting doors and windows.

Now that you have sketched the walls, you will add some doors. A door symbol indicates which side the hinges are on and the direction the door opens.

New doors are typically shown open 90 degrees in floor plans which helps to avoid conflicts such as the door hitting an adjacent base cabinet. Additionally, to make it clear graphically, existing doors are shown open only 45 degrees.

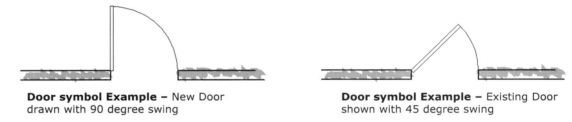

Door symbol Example – New Door
drawn with 90 degree swing

Door symbol Example – Existing Door
shown with 45 degree swing

One of the most powerful features of any CAD program is its ability to reuse previously drawn content. With Revit you can insert entire door systems (door, frame, trim, etc). You drew a 2D door symbol in Lesson 4; however, you will use Revit's powerful door tools which are fully 3D.

For those new to drafting, you may find this comparison between CAD and hand-drafting interesting: when hand-drafting, one uses straight edges or a plastic template that has doors and other often used symbols to trace.

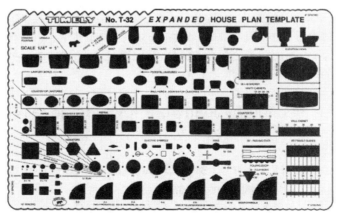

Hand Drafting Template – Plastic template with holes representing common residential shapes (at ¼" – 1'-0")
Image used by permission: Timely Templates www.timelytemplates.com

Doors in stud walls are dimensioned to the center of the door opening. On the other hand, and similar to dimensioning masonry walls, doors in masonry walls are dimensioned to the face (see example below).

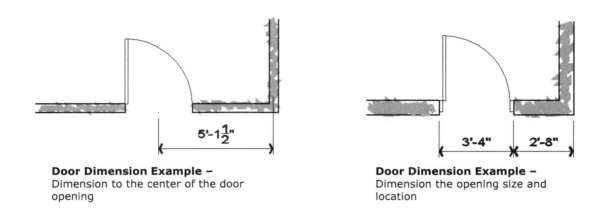

Door Dimension Example –
Dimension to the center of the door opening

Door Dimension Example –
Dimension the opening size and location

Loading additional door families:

Revit has done an excellent job providing several different door libraries. This makes sense seeing as doors are an important part of an architectural project. Some of the provided libraries include bi-fold, double, pocket, sectional (garage), and vertical rolling, to name a few. In addition to the library groups found on your local hard drive, many more are available via the Web Library feature (which requires an Internet connection).

The default template you started with (Residential-Default.rte) only provides four door *Families* (each family contains multiple door sizes). If you want to insert other styles you will need to load additional *Families* from the library. The reason for this step is that, when you load a family, Revit actually copies the data into your project file. If every possible group was loaded into your project at the beginning, not only would it be hard to find what you want in a large list of doors, but also the files would be several megabites in size before you even drew the first wall.

You will begin this section by loading a few additional *Families* into your project.

1.   Open project ex5-3.rvt and **Save-As ex5-4.rvt**.

2.   With the *Door* tool selected, select **Load...** on the *Options Bar*. (Figure 5-4.1)

3. Browse through the **Doors** folder for a moment.
(The Doors folder is a sub-folder of *Imperial Library*; Revit should have taken you there by default. If not, you can browse to C:\Documents and Settings\All Users\Application Data\Autodesk\Revit Building 8\Imperial Library\Doors.)

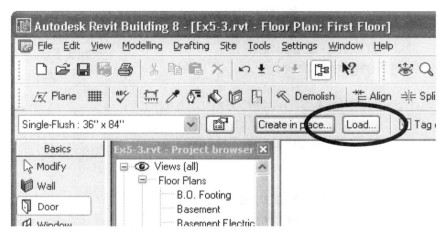

Figure 5-4.1 Load from Library

Each file represents a *Family*, each having one or more similar elements (of varying size). Next you will load four door *Families* into your project.

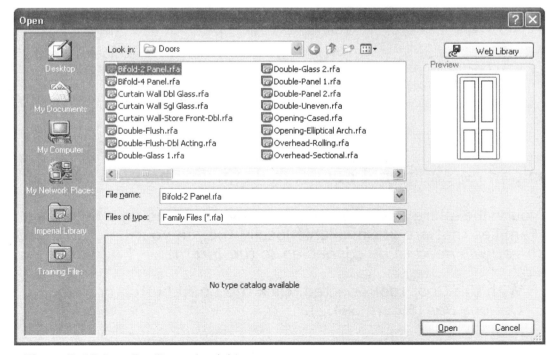

Figure 5-4.2 Door Families on hard drive

4. Select **Bifold-2 Panel.rfa**, and then click **Open** (Figure 5-4.2).

As soon as you click *Open*, the selected *Family* is copied into your current project and is available for placement.

5. Repeat steps 2–4 to load the following door groups:
    a. **Opening-Elliptical Arch**
    b. **Sliding-2 panel**
    c. **Single-Panel 4**

6. In the *Project Browser*, expand the *Families* category and then *Doors* to see the loaded door *Families*. (Figure 5-4.3)

If you expand the *Doors* sub-category itself in the *Project Browser*, you see the predefined door sizes associated with that group. Right-clicking on a door size allows you to rename, delete or duplicate it. To add a door size you duplicate and then modify properties for the new item.

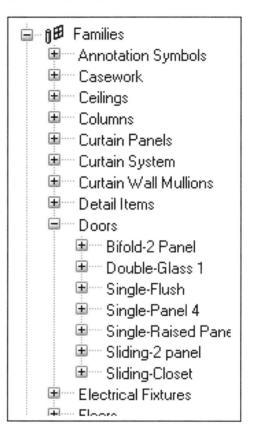

Figure 5-4.3 Loaded door families

Downloading additional door families from the Web Library:

Now you will explore Revit's Web Library by downloading a few more door Families that are not available locally (i.e. on your hard drive). *Of course you will need to be connected to the Internet.*

7. With the *Door* tool selected, click the **Load** button on the *Options Bar* (Figure 5-4.1).

8. Click the **Web Library** button in the upper right (Figure 5-4.2).

Revit will open your web browser, then you will be looking at the contents of Revit's Online Content Distribution Center. (Figure 5-4.4)

9. Click on "**Revit 8.0 Library**".

Figure 5-4.4 Web content via browser

Figure 5-4.5 Web content categories

10. Click the **Doors** category (Figure 5-4.5).

Within the web browser you should see a graphical representation of each family available for download. (Figure 5-4.6)

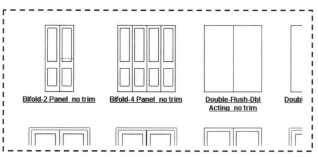

Figure 5-4.6 Web content visuals

11. Click "**Double-Raised Panel with Sidelights**" to download that family.

12. Select **Save**. (Figure 5-4.7)

13. *Save* the file to your computer's *Desktop* using the default name provided.

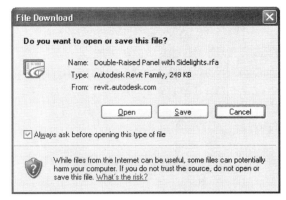

Figure 5-4.7 Web browser download prompt

14. Repeat the previous steps to download three more door families from the Web Library:
    a. **Overhead-Sectional-Flush**
    b. **Single-Entry 3**
    c. **Single-Pocket**

Now that you have saved the door family file to the hard drive, you need to load it into your current project. Unfortunately, this process is not as automated as loading local content.

You may want to create a folder where you save all the *Families* you download. This would make future access more convenient, especially if you did not have internet access at that time. This folder should probably not be in Revit's standard folder location as it may be difficult to migrate the data after an upgrade (i.e. sorting the custom files from the default files).

One more thing on *Families* via the web: you have access to content via sources other than Revit. You can try a Google search with text that reads something like "*autodesk revit families*". As Revit's popularity grows, product manufactures will start making *Families* available that represent their products (thus making it easier for designers to incorporate their products into a project). Currently, Anderson Windows and the Woodwork Institute (cabinets) have created Revit content, both of which are available at Revit's Web Library. Additionally, you can visit www.augi.com (Autodesk Users Group International) or www.revitcity.com for Revit related resources and content.

15. Select **File → Load From Library → Load Family...** from the menu bar.

16. Browse to the *Desktop* and select **Double-Raised Panel with Sidelights.RFA**, and then click **Open** (similar to Figure 5-4.2).

17. Repeat the previous two steps to load the other three *Families* (from step 14) into your project.

18. In the *Project Browser*, expand the Doors category again to verify all the doors have been loaded into your project. (Figure 5-4.8)

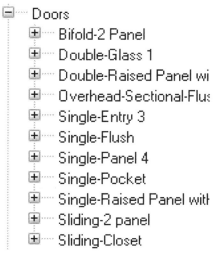

Note that the *Opening-Elliptical Arch* does not display here; it actually shows up under *Generic Model*.

Figure 5-4.8 All door families loaded for this exercise

Next you will start inserting the doors into the first floor plan.

19. With the *Door* tool selected, pick **Single-Entry 3: 36" x 80"** from the *type selector* on the *Options Bar*.

20. Insert one door in the north wall of the garage as shown in Figure 5-4.9.

Immediately after the door is inserted, or when the door is selected, you can click on the horizontal or vertical flip-controls to position the door as desired.

Revit automatically adds a door tag near the newly inserted door. The door itself stores information about the door that can then be compiled in a door schedule. The doors are automatically numbered in the order they are placed. The number can be changed at any time to suit the designer's needs (you will do this later in the door schedule section). Most architects/designers use 1x or 1xx (i.e. 12 or 120) numbers for the first floor and 2x or 2xx numbers for the second floor.

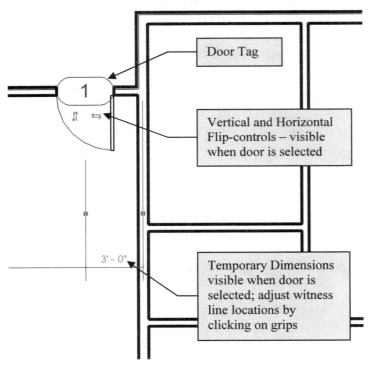

The door tag can be moved simply by clicking on it to select it, and then clicking and dragging it to its new location.

Also, with the Door Tag selected, you can set the vertical/horizontal orientation and if a leader should be displayed when the tag is moved; it is done via the Properties button on the *Options Bar*.

Figure 5-4.9 First door placed at rear of garage

21. With the newly inserted door selected, click the *Properties* button on the *Options Bar*.

Notice, in Figure 5-4.10, the upper pane shows the *Type Parameters* which are common to all "Single-Entry 3" doors in the Project. The lower pane shows the *Instance Parameters* which are options that vary with each door.

The *Type Parameters* are not editable in this dialog. To be sure you want to make changes that affect all doors of that type, Revit forces you to click the *Edit/New...* button and make changes in the *Type Properties* dialog. Also, in the Type Properties dialog, you can add additional

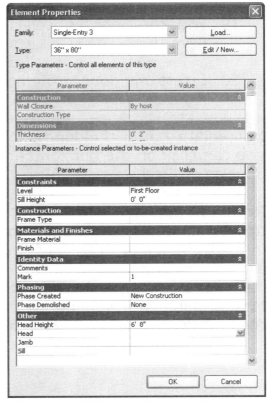

Figure 5-4.10 Element Properties for door

pre-defined door sizes to the "Single-Entry 3" door family by clicking the Duplicate button, providing a new name and changing the "size" dimensions for that new Type.

22. Click **OK** to close the *Element Properties* dialog.

23. Finish inserting doors for the first floor. (Figure 5-4.11) Use the following guidelines:

   a. Use the door style shown in Figure 5-4.11.
   b. Doors across from each other in the Mud Room should align with each other. ***TIP***: *While inserting the second set of doors, watch/wait for the reference line to show up, indicating alignment.*
   c. Place doors approximately as shown when exact location not given.
   d. See the next page for information on sliding pocket doors.
   e. While inserting doors, you can control the door swing by moving the cursor to one side of the wall or the other and by using the space bar on the keyboard.
   f. The order in which the doors are placed does not matter; this means your doors may have different numbers than the ones shown below.

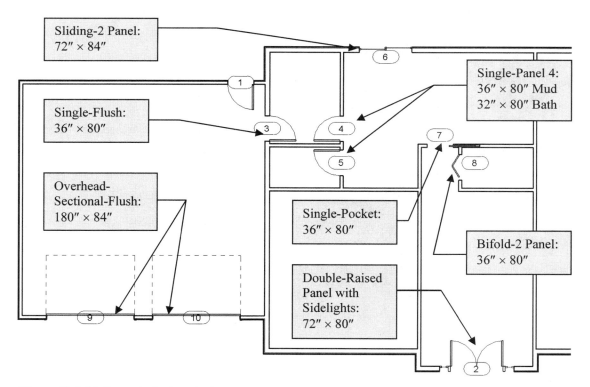

Figure 5-4.11 Doors to be inserted

Sliding pocket doors:

Sliding pocket doors are doors that slide into the wall rather than swinging open into a room. This type of door is convenient when the door will be open most of the time and you do not want, or have room for, an open door. On the other hand, if the door will be used often, like the door from the hallway into the bath room, you should use a swinging door for ease of use and long-term durability.

One important thing to keep in mind when designing a room with a sliding door: the entire width of the door is in the wall when the door is fully open. That means you need space in the wall next to the door that equals the width of the door plus a few inches depending on the manufacturer (this includes plumbing and electrical that might be in the wall).

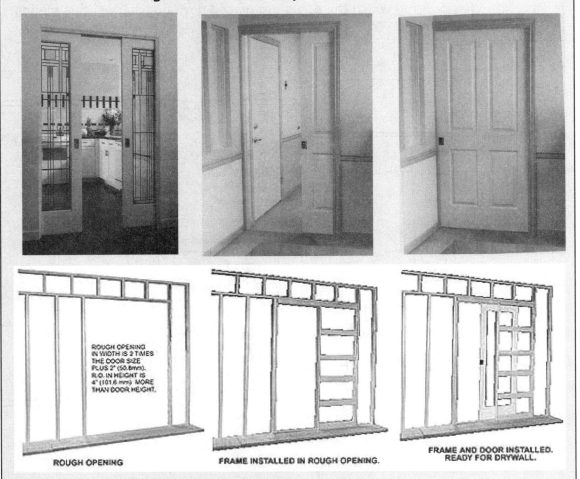

POCKET DOOR EXAMPLES – Double door and Single door photos (top);
Installation illustration (bottom).
Images used by permission: KrisTrack www.kristrack.com

Insert Openings:

With the first floor doors in place, you will add three openings. An opening is similar to a door in that, when placed, it removes a portion of wall automatically creating a circulation route. Also, like doors, openings come in different pre-defined shapes and sizes.

One common mistake is to break a wall so the floor plan appears to have an opening. However, when a section is cut through the opening there is no wall above the opening. This is also problematic for the room tagging feature and the Ceiling placement tool.

One of the keys to working successfully in Revit is to model the building project just like you would build it because that is the way Revit is designed to work.

Next you will place two larger openings in the walls between the Foyer and the Living room and the Dining Room. A smaller opening will be placed in the wall between the Dining Room and the Kitchen. All three openings are from a family you previously loaded: *Opening-Elliptical Arch*.

The Type Properties dialog is shown below (Figure 5-4.12). Notice the various fields that can be filled in. For example, you could enter how much it would cost to construct an opening with an arched top in the Cost field.

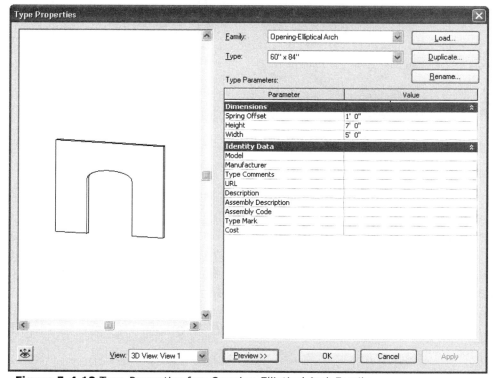

Figure 5-4.12 Type Properties for: *Opening-Elliptical Arch* Family

24. Select the **Component** tool, under the *Basics* tab, on the *Design Bar*.

25. Place the three **Openings** as shown in Figure 5-4.13.
    TIP: *Select the appropriate opening (family/type) from the Type Selector on the Options Bar.*

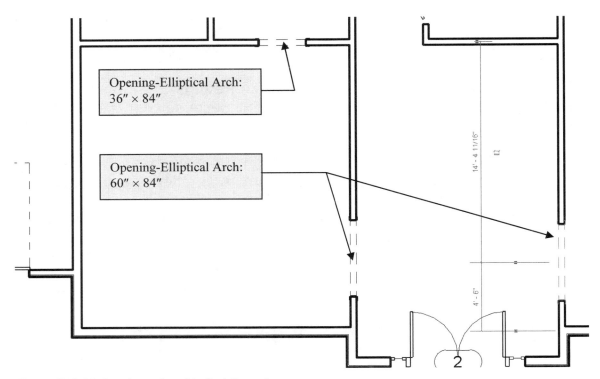

Figure 5-4.13 Openings placed in first floor plan

Insert Windows:

Adding windows to your project is very similar to adding doors. The template file you started from has three Families preloaded into your project: Casement with Trim, Double Hung with Trim and Fixed with Trim. Looking at the *Type Selector* drop-down you will see the various sizes available for insertion. Additional Families can be loaded (similar to how you loaded the doors), although your project will only use two Families total and they are already loaded (Double Hung and Fixed).

1. Using the **Window** tool, on the *Basics* tab, select **Double Hung with Trim: 36" x 72"** in the *Type Selector*.

2. Insert the windows as shown in **Figure 5-4.14**; be sure to adjust the window locations to match the dimensions shown (do not add the dimensions).

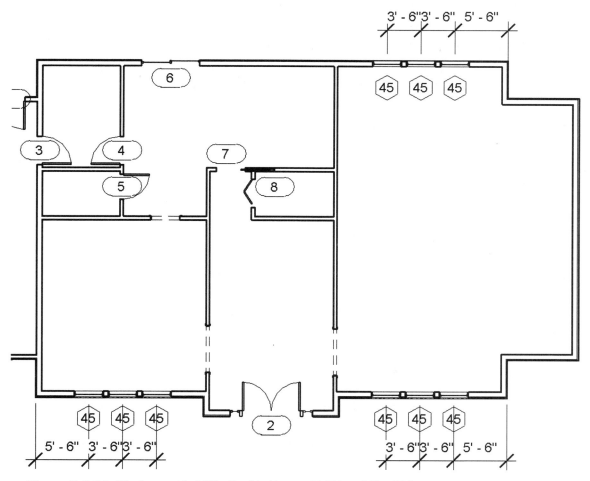

Figure 5-4.14 Windows added (9): Double Hung with Trim: 36" x 72"

With a window type within a *Family*, not only is the window size predefined, but the vertical distance off the ground is as well. Next you will see where this setting is stored.

3. Select one of the windows that you just placed in the walls and then select **Properties** from the *Options Bar*.

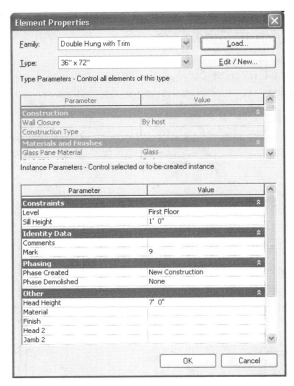

Figure 5-4.15 Element Properties

As previously discussed, the lower portion of this dialog shows you the various settings for the selected item (a window in this case). See Figure 5-4.15.

As you might guess, the same size window could have more than one sill height (i.e. distance from the floor to the bottom of the window). So Revit lets you adjust this setting window by window.

Another setting you might have noticed by now is the *Phase Created*. This is part of what makes Revit a 4D program (the forth dimension is time). Again, you might have several windows that are existing and 2-3 new ones in an addition; all the same size window. Revit lets you do that!

4. Make sure the *Sill Height* is **1'-0"**. (If not, change all 9 windows.)

5. Click **OK** to close the dialog.

Next you will insert several double hung windows that are not as tall, in the less prominent rooms; also in the kitchen so cabinets will fit below the window. However, if you did place the tall windows in the kitchen you would realize it as soon as you tried to place the base cabinets and generate interior elevations (or sections) in the kitchen. This is because the entire Revit project is one 3D database.

Compare the above mentioned conflict with traditional CAD, where the 2D elevations have nothing to do with the floor plans, sections, or interior elevations. With traditional CAD, you could draw the exterior elevations with the tall windows and place cabinets under it in plan view and then draw an interior elevation showing the cabinets and a shorter window. The windows are usually ordered and rough openings framed based on the exterior elevations. Revit helps to avoid these types of costly mistakes.

6. Insert window **Double Hung with Trim: 36" x 48"** as shown in **Figure 5-4.16**; use *Properties* to make sure the *Sill Height* is set to **3'-0"**.

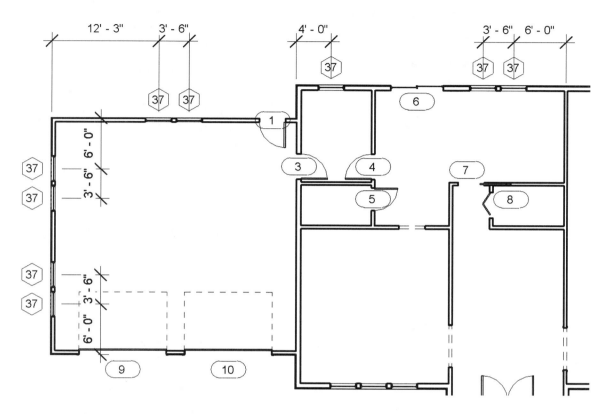

Figure 5-4.16 Windows added (9): Double Hung with Trim: 36" x 48"

Creating a new Window Type in an existing Family:

You will add four more windows to the east wall. These will be high windows; two on either side of a fireplace. Rather than using one of the predefined window sizes, you will create a new size within an existing Family.

These windows will be fixed (non-operable), so you will be adding a new type to the Fixed with Trim *Family*.

7. Select the **Window** tool and then select the **Properties** button on the *Options Bar.*

8. Set the *Family* drop-down to **Fixed with Trim** and then set the *Type* drop-down to **36" x 24"** (see Figure 5-4.17).

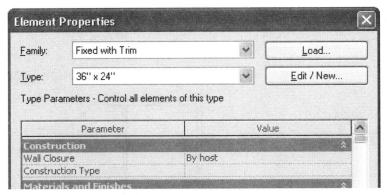

Figure 5-4.17 Element Properties –
Family: Fixed with Trim; *Type*: 36" x 24"

9. Click **Edit / New...**

10. Click the **Duplicate** button.

11. Enter **36" x 20"** for the name and then click **OK**.

You are now viewing the *Type Properties* for the new 36" x 20" type you just created. However, the name has nothing to do with actually changing the window size. You will do this next.

12. Change the Height Parameter to 1'-8".

13. Change the Default Sill Height to 5'-4. *(This makes the top of all your windows align at 7'-0".)*

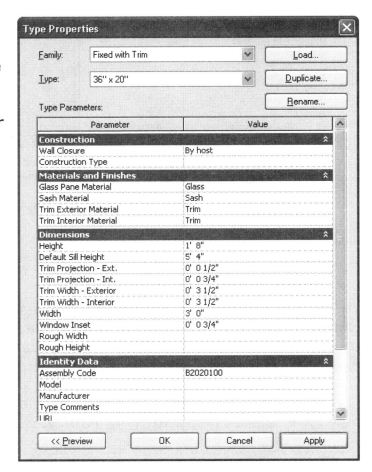

Figure 5-4.18 Type Properties: New pre-defined window size

14. Click **OK** twice to close the dialogs.

15. Add the windows as dimensioned in Figure 5-4.19.

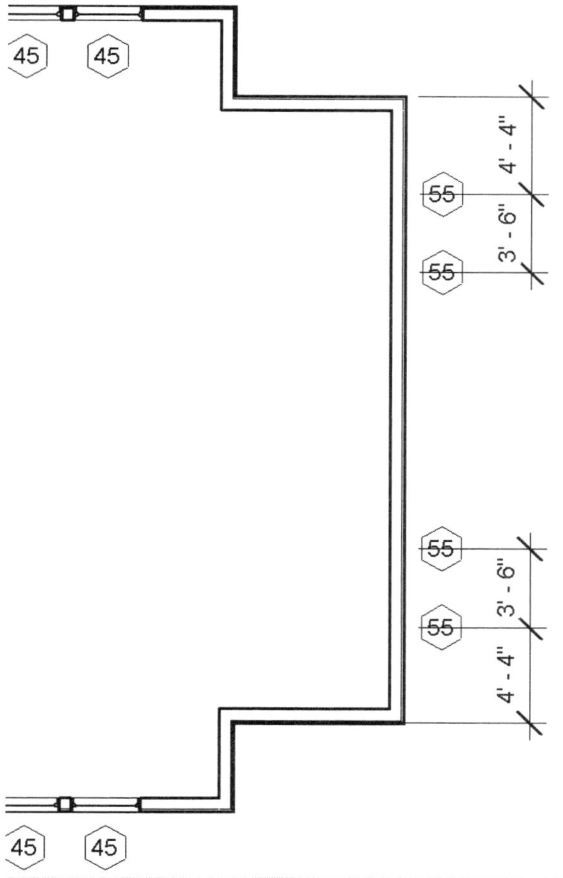

Figure 5-4.19 First floor plan: last four windows added

You are probably wondering where the windows are. They are tagged and you can select them. However, they are not visible.

With your project being a 3D model, Revit has to cut through the model at a certain elevation when generating the plan view. The default Cut Plane is 4'-0" above the floor line.

As you should recall, the sill of the new window type you just created is 5'-4" (Figure 5-4.18).

So the window occurs above the cut line. The windows will show up just fine in elevation and section views. Next you will see where the Cut Plane is set for each view.

16. Right-click anywhere in the drawing window (of the first floor plan view) and select **View Properties...** from the pop-up menu.

17. Scroll down and select **Edit** next to *View Range*.

Notice the *Cut Plane* is set to 4'-0" Figure 5-4.20. Try changing it and closing the dialogs to see the various effects it has on your floor plan view (look at the stairs and the high windows).

Be sure to set the Cut Plane back to 4'-0" before proceeding.

18. Click **OK** to close the open dialog boxes.

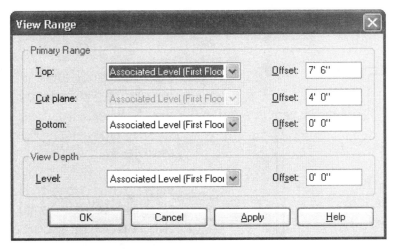

Figure 5-4.20 View Range properties

Cleaning house:

As previously mentioned, you can view the various Families and Types loaded into your project. The more Families and Types you have loaded, the larger your file is whether or not you are using them in your project. Therefore, it is a good idea to get rid of any door, window, etc., that you know you will not need in the current project. You do not have to delete any at this time, but this is how you do it:

- In the *Project Browser*, navigate to Families → Windows → Fixed. Right-click on **36" x 48"** and select **Delete**.

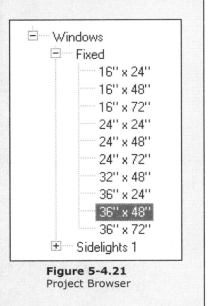

Figure 5-4.21
Project Browser

Using the Array command:

The *Array* tool allows you to quickly copy several objects that have the same distance between them. This command is not essential to adding the windows but can be used on the grouping of three windows, if you want to try using *Array*.

- Click the *Modify* tool and then select an element (an exterior window in this example).

- With the window selected, pick the **Array** tool from the *Edit* toolbar.

- In the *Options Bar*, type **6** for the *Number of Items* field.

- Click the left mouse button at the midpoint of the window and move your mouse to the right until the dimension displayed is 8'-6".

- You should now see the windows arrayed in the wall. 8'-6" is not coursing, so select the **Activate Dimensions** button on the *Options Bar* and then enter 8'-8" in the displayed dimension to adjust the window openings. This allows you to more accurately adjust the dimensions.

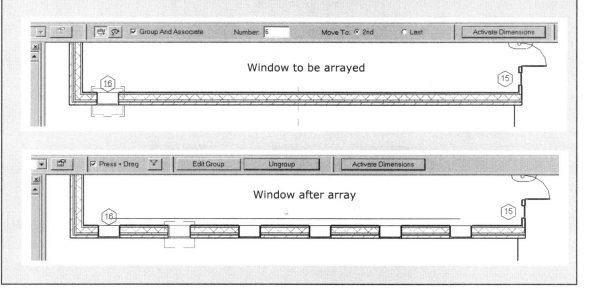

That concludes the exercise on adding doors, openings and windows to the first floor plan.

Exercise 5-5:
Adding a fireplace

The last thing you will do in this lesson is to add a fireplace to the Living Room. Creating a fireplace is beyond the scope of this introductory book, so you will download one from the Web Library.

1. Using techniques previously described to download doors from the *Web Library* (page 5-39), download the following items (Figure 5-5.1):
    a. Revit 8.0 Library → Misc. Architectural → **Mantel 1**
    b. Revit 8.0 Library → Misc. Architectural → **Masonry Chimney-Wall**

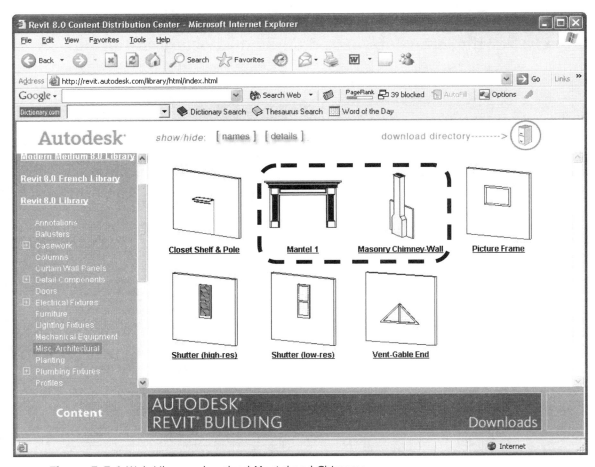

Figure 5-5.1 Web Library: download Mantel and Chimney

2. Load the downloaded *Families* into your project (see page 5-41).

You are now ready to place the fireplace in the first floor plan.

3.   With the <u>First Floor Plan</u> *View* current, select **Component** from the *Basics* tab.

4.   From the *Type Selector*, pick **Masonry Chimney-Wall**.

The chimney will be looking for a wall to be inserted into. When your cursor is near a wall you will see a "ghost" chimney to help you visually place the element. When the cursor is not near a wall, Revit does not display the "ghost" image because it is not a valid location to place the selected element.

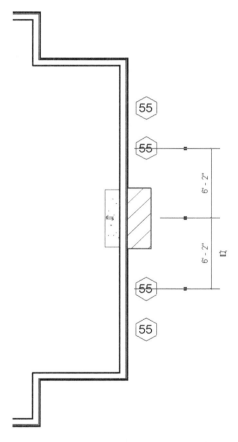

Figure 5-5.2 Chimney added

Try moving the cursor around the plan for a minute. Notice, again, that when the cursor is near a wall you see a "ghost" image and the image orientates itself with the wall (e.g. vertical, horizontal, angled). When the element encounters another element placed within a wall, you will see a small circle with a diagonal line through it indicating the two elements cannot overlap. Furthermore, the element flips to the side of the wall the cursor is on. Try moving the cursor from one side of the wall to the other and observe the image flip back and forth.

5.   Move your cursor to the middle portion of the far east wall (between the high windows). With your cursor on the exterior side of the wall, click to place the fireplace.

6.   Adjust the temporary dimensions so the chimney is centered on the wall. (Figure 5-5.2)

Next you will place the mantel in the plan view. This is less automatic as it does not attach to the wall so you will insert the mantel in the middle of the living room, rotate it and then move it into place.

7. Use the Component tool to insert Mantel 1:80" x 58 ¾" into the middle of the living room.

8. Select the Mantel and use the **Rotate** tool to rotate the mantel 90 degrees in the clockwise direction.

9. With the Mantel still selected, use the arrow keys on your keyboard to "nudge" the element into place. (Figure 5-5.3)

10. Take a minute to look at the 3D View of all that you have completed so far (Figure 5-5.4 and image below).

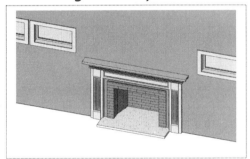

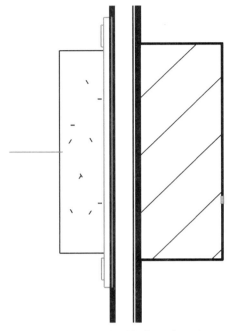

Figure 5-5.3 Mantel added in plan view

11. **Save** your project as **Ex5-5.rvt**.

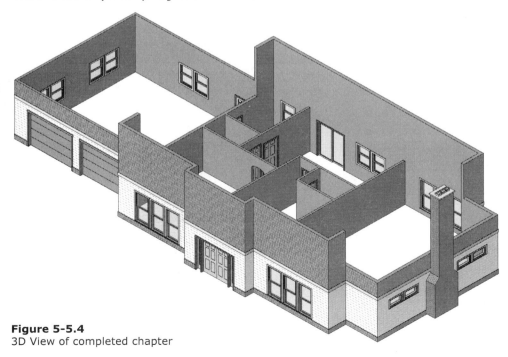

Figure 5-5.4
3D View of completed chapter

Self-Exam:

The following questions can be used as a way to check your knowledge of this lesson. The answers can be found at the bottom of this page.

1. The Option Bar allows you to select the height your wall will be drawn at. (T/F)

2. It is not possible to draw a wall with the interior or exterior face of the core as the reference point. (T/F)

3. Objects cannot be moved accurately with the Move tool. (T/F)

4. The _____ tool, in the Design Bar, has to be selected in order to select an object in your project.

5. A wall has to be _____ to see its flip icons.

Review Questions:

The following questions may be assigned by your instructor as a way to assess your knowledge of this section. Your instructor has the answers to the review questions.

1. Revit comes with many predefined doors and windows. (T/F)

2. The Project Information command allows you to enter data about the project, which is automatically added to sheet borders. (T/F)

3. You can delete unused families and types in the Project Browser. (T/F)

4. It is not possible to add a new window type to a window family. (T/F)

5. It is not possible to select which side of the wall a window should be on while you are inserting the window. (T/F)

6. What tool will break a wall into two smaller pieces? _____

7. The _____ tool allows you to match the surface of two adjacent walls.

8. Use the _____ key, on the keyboard, to flip the door swing while placing doors using the Door tool.

9. You adjust the location of a dimensions witness line by clicking on (or dragging) its _____ .

10. The _____ file has a few doors, windows and walls preloaded in it.

Self-Exam Answers:
1 – T, **2** – F, **3** – F, **4** – Modify, **5** – Selected

Notes:

Lesson 6
Residence: SECOND FLOOR & BASEMENT PLANS

In this lesson you will set up the second floor plan as well as the basement. This will involve copying objects from the first floor with some modifications along the way. You will also add stairs and dimensions to your plans.

Exercise 6-1:
View setup and enclosing the shell

Discussion: Second Floor View current conditions

1. Switch to the **Second Floor** view under *Floor Plans* in the *Project Browser* (double-click on the view name).

You should see a plan view similar to Figure 6-1.1. The geometry you are seeing is a result of two things:

- First, the height of walls drawn on the first floor:
    i.   Some of the exterior walls that you sketched in the first floor plan are 19'-0" tall.
    ii. From the elevation views, you can see that the second floor is 9'-0" above the first.
    iii. As you may recall from the last chapter, the default View Range – Cut Line is 4'-0" above the floor.
    iv. So, relative to the first floor, the second floor Cut Plane is 13'-0" above it (9'-0" + 4'-0").
    v.   Therefore, the 19'-0" tall walls extend past the second floor Cut Plane; that is why they are bold (elements in section automatically have a heavier lineweight).
    vi. The exterior walls of the garage and east living room walls are only 12'-0", which is below the Cut Plane. It is still within the View Range so it shows up in elevation (so to speak). Elements in elevation have lighter lines than those in section.

- Second, is the Underlay feature:
    i.   In the View Properties for the second floor plan view you will see a parameter labeled Underlay. (Figure 6-1.2)
    ii. This feature is currently set to First Floor.

iii. The Underlay feature causes the selected view's walls to be displayed as light gray lines for reference.

iv. The walls shown as part of an underlay cannot be modified.

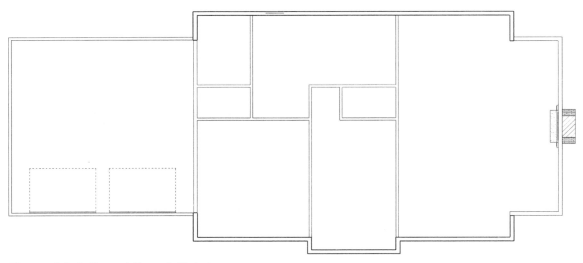

Figure 6-1.1 Second Floor: initial view

You do not need to see the first floor walls at the moment, so you will turn them off.

2. In the *View Properties* dialog, set **Underlay** to **None**.
   (Right-click in the drawing window and select View Properties from pop-up menu.)

3. Click **OK**.

Now you should only see the exterior walls. (Figure 6-1.3)

Next you will sketch two more exterior walls to enclose the second floor. These walls only occur above the second floor line so they will only have wood shakes for siding.

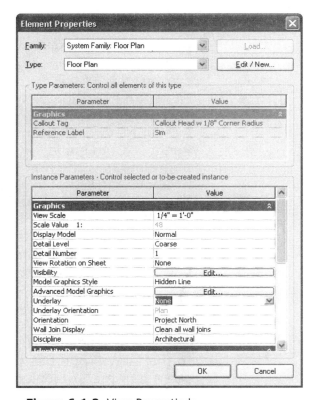

Figure 6-1.2 View Propertied: Second Floor view

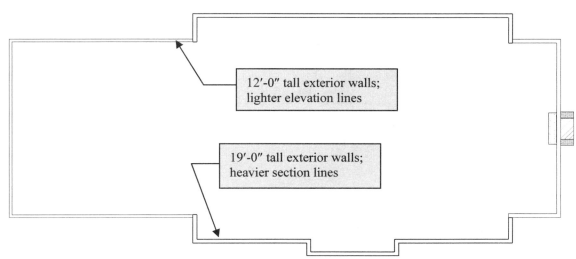

12'-0" tall exterior walls; lighter elevation lines

19'-0" tall exterior walls; heavier section lines

Figure 6-1.3 Second Floor: underlay turned off

4. Using the *Wall* tool, select **Basic Wall: Exterior – Wood Shingle on Wood Stud** from the *Type Selector*.

If you look at the properties for this wall you will see that it is very similar to your primary exterior wall assembly. The main difference is that the vertical elements are stripped away (horizontal siding and two sweeps). You will see another wall type with just horizontal siding as well. These two walls are loaded in the template for the reason you are about to use one as infill between the main (more complex) wall type.

If you tried to use the wall type "Basic Wall: Exterior – Wood Shingle over Wood Siding on Wood Stud", the horizontal wood siding would start at the second floor line for any wall placed on the second floor. This is why you need a wall that just has wood shingles and no sweeps (i.e. horizontal trim boards).

5. Draw the two walls as shown in Figure 6-1.4, using the following information:
    *   Set the *Height* to **Roof** (*Options Bar*).
    *   Set the *Loc Line* to **Finished Face: Interior** (*Options Bar*).
    *   Pick the interior side of the existing wall (press the Space Bar if you need to flip the wall) before picking the wall end point.

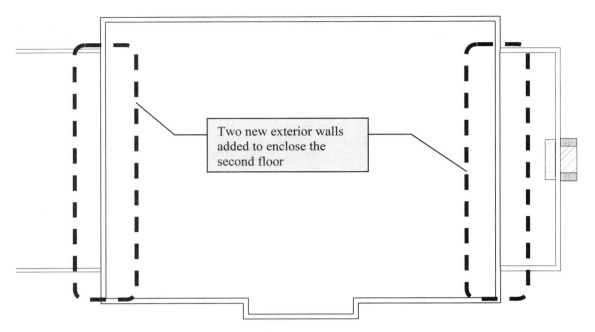

Figure 6-1.4 Second Floor: two exterior walls added

The height of the walls will change later when the roof is drawn. In fact, the height will vary as it extends up to the underside of the gable end roof (see image on book cover).

The two walls just drawn will be supported by the second floor, which has not been drawn yet.

6. **Save** your project as **Ex6-1.rvt**.

FYI:

You have probably figured out on your own what the symbol at the right is for. If not, here it is:

The default template has four Elevation symbols shown in plan view. These symbols represent what the four pre-setup views (under elevation) will see. Therefore, you should start drawing your plan in the approximate center of the four symbols. The symbols can be moved by dragging them with your mouse. This is covered more thoroughly later in the book.

Exercise 6-2:
Adding the interior walls

This short exercise will help reinforce the commands you have already learned. You will add walls, doors and windows to your project.

Copying walls from the first floor:

Often, you will have walls occur in the same location on each floor. The easiest thing to do is copy/paste between floors. You will try this next.

1. Switch to the **First Floor** plan view.

2. Hold down the **Ctrl** key, select the four walls identified in Figure 6-2.1.

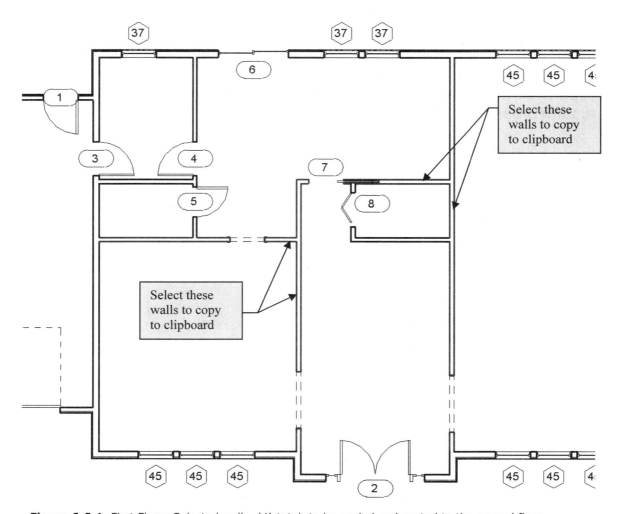

Figure 6-2.1 First Floor: Selected walls, (4) total, to be copied and pasted to the second floor

3. With the four walls selected, select **Copy to Clipboard** from the *Edit* pull-down menu.

4. Switch back to the **Second Floor** plan view.

5. From the *Edit* pull-down menu, select **Paste Aligned → Current View**. (Figure 6-2.2)

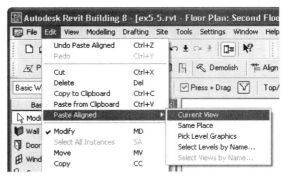

You should now see the four walls in your second floor plan view (Figure 6-2.2).

Figure 6-2.2 Edit Menu: Paste Aligned fly-out

Why not draw these walls 19'-0" high like the exterior walls?

Simulating real-world construction is ideal for several reasons. Mostly, you can be sure exterior walls align from floor to floor. Although the four walls align, they do not necessarily have to because they are separated by floor construction, this allows one floor to be later modified easily.

6. Select **Modify** to deselect the walls.

Any time a wall is copied, or copy/pasted, any elements attached to the wall(s) is copied with the wall. Next you will delete the openings that were copied with the walls from the first floor.

7. Individually select each of the four openings (door / opening) and delete them (delete key on keyboard).

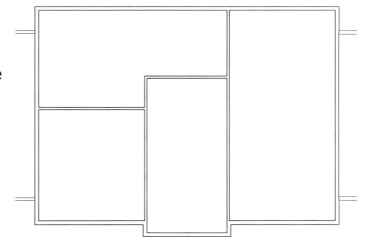

Figure 6-2.3 Second Floor: walls copied, doors deleted

Your first floor walls have their *Top Constraint* set to the second floor, so when you copied the walls from the first floor to the second floor Revit automatically set the *Top Constraint* to the next level above the first floor; the roof level. This is fine for now.

Next you will add a few more interior walls.

8. Using the **Wall** tool, draw the walls shown in Figure 6-2.4 using the following information:
    a. Wall type: **Basic Wall: Interior – 4 1/2" Partition**
    b. Loc Line: *use whatever is appropriate*
    c. Height: **Roof**
    d. Do not draw the dimensions at this time.
    e. Notice that one of the interior walls from Figure 6-2.3 gets extended and another gets trimmed; the walls previously created have a heavy dashed line over them.
        (Figure 6-2.4)

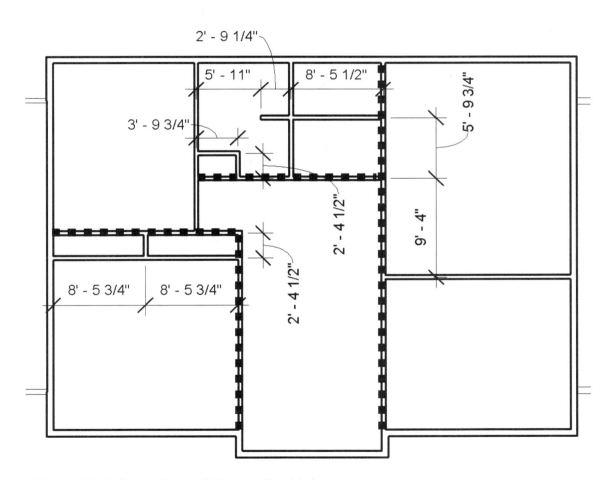

Figure 6-2.4 Second Floor: additional walls added

Joining Geometry:

Occasionally the intersection of two walls will not clean up, so you can use the *Join Geometry* icon on the Tools toolbar. The steps are as follows:

- Click the *Join Geometry* icon.

- Select the two walls.

- That's it; the walls are joined.

Intersecting walls

Walls joined

Controlling the View Scale:

The size of the dimensions and door and window tags are controlled by the *View Scale*. Each *View* has its own scale setting. To facilitate the creation of this book, the *View Scale* has been set to 3/32" = 1'-0" typical. At that scale, Revit makes the text and symbols larger so it is legible at the smaller scale (however, the images in this book are not to scale).

The default *View Scale* is ¼" = 1'-0" for the plan views (i.e. floor, ceiling and framing plans). This is good and should not be changed for this tutorial. In addition to controlling the text and symbols, it also controls the size (or scale) of that drawing on the plot sheet (i.e. one of the sheets in a set of drawings). This only makes sense seeing as *View Scale* has everything to do with what scale that view is intended to be printed at.

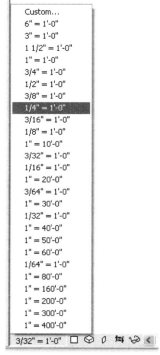

Figure 6-2.5 View Scale

The *View Scale* can be controlled via the *View Properties* dialog or more conveniently on the *View Control* bar at the bottom of the *Drawing Window* (Figure 6-2.5).

9. Save your project as **ex6-2.rvt**.

Exercise 6-3:
Adding Doors, Openings & Windows

Don't forget to keep a backup of your files on a separate disk (i.e., floppy, CD or DVD). Your project file should be about 7.5 MB when starting this exercise. At 1.4 MB the file will not fit on a floppy disk. That should not be a problem as most computers today have a CD Burner or a ZIP drive you can use to save large files to. Remember, your Revit project is one large file (not many small files). You do not want anything to happen to it!

Adding more doors:
Now you will add the second floor doors and openings, but first you must load one more door Family into your project.

1. Open project ex6-2.rvt and **Save-As ex6-3.rvt**.

2. Load the door Family **Bifold–4 Panel.rfa** into your project.
    Reminder: *Click the Door tool, Load button (Options Bar), and then browse to the Doors sub-folder.*

3. Add the doors as shown in Figure 6-3.1
    a. Center the closet doors in the closet.
       TIP: *With the inserted door selected, you can "nudge" the door using the arrow keys FYI: the more zoomed in you are, the more accurate the "nudge" increments are.*

Residential Door sizes!
Doors come in many shapes and sizes.

Thickness: 1 3/8", 1 3/4" and 2 1/4"
The 1 3/8" door is most often used in residential construction and the 2 ¼" door is typically only used on exterior doors for high-end projects – i.e. big budget.

Size: ranges from 1'-0" x 6'-8" to 4'-0" x 9'-0"
The height of a residential door is usually 6'-8". This writer recently worked on a project where 7'-0" high doors were specified. The contractor was so used to the standard 6'-8" dimension; he framed all the exterior openings at 6'-8". To top it off, the exterior walls where concrete (ICF system), so the openings could not be modified. Thus, 6'-8" doors were used.

You can find more information on the internet; two examples are:
      algomahardwoods.com   and   madawaska-doors.com
      (wood veneer door mfr.)   *(custom solid wood door mfr.)*

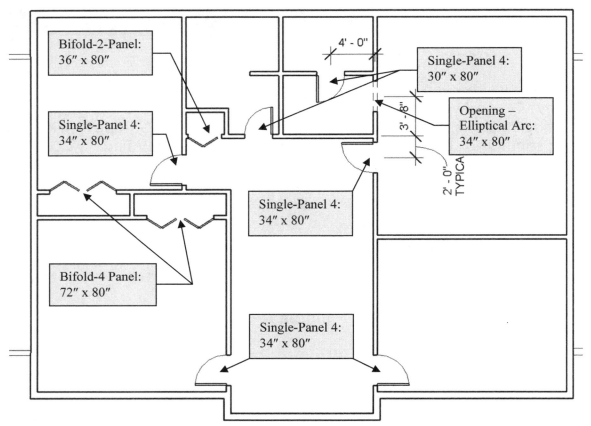

Figure 6-3.1 Second Floor; doors added

The door numbers were temporarily hidden from view for clarity. On your plans you notice that the new doors have different numbers while the windows have the same number. Why is this? It relates to industry standards for architectural drafting. Each window that is the same size and configuration has the same type number through-out the project. Each door has a unique number because doors have so many variables (i.e. locks, hinges, closer, panics, material, and fire rating). To make doors easier to find, many architectural firms will make the door number the same as the room number the door opens into. You can change the door number by selecting the symbol and then clicking on the text. The door schedule will be updated automatically.

Adding more windows:

4. Add the windows as shown in Figure 6-3.2.
    a. Verify via *Properties* that each window's *Head Height* is set to 7'-0".   ***TIP**: Select the window and click the Properties button.*
    b. All windows are **Double Hung with Trim: 36" x 48"** unless otherwise noted in Figure 6-3.2.
    c. Similar to the previous image, the door and window tags have been temporarily hidden from view (you will learn how to do this next).
    d. All dimensions are from the center of the window to the exterior face of wall.

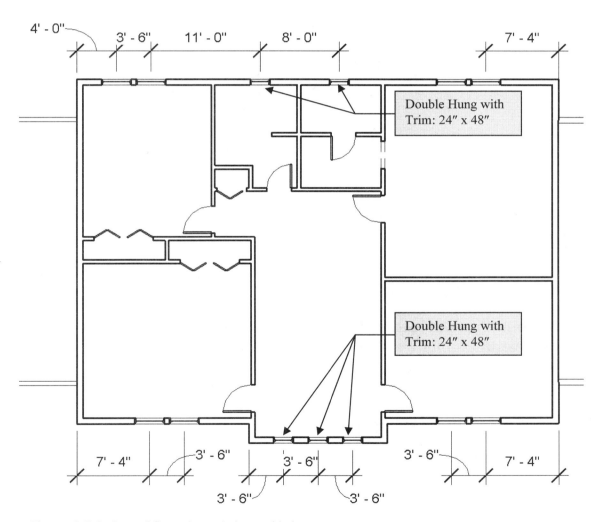

Figure 6-3.2 Second floor plan: windows added

Using Hide/Isolate to control visibility:

Now you will shift gears for a moment and look at the *Hide/Isolate* feature. This allows you to temporarily hide or isolate the selected elements or all the elements in the same category as those selected.

This feature is accessed via the *View Control* bar, similar to the *View Scale* feature previously discussed.

5. Holding the **Ctrl** Key, select one door tag and one window tag.

6. One the *View Control* bar click the **Hide/Isolate** icon (looks like sunglasses).

7. From the pop-up menu, select **Hide Category**. (Figure 6-3.3)

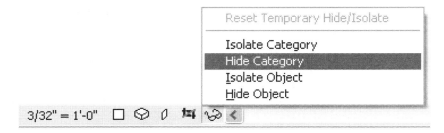

Figure 6-3.3 View Control bar: Hide/Isolate selected

You should notice two things. First, all the door and window tags are gone. Second, the Hide/Isolate icon has a red background.

As implied by the feature name, the tags are simply hidden from view; they have not been deleted.

If you would have selected Hide Object, only the selected elements would have been hidden from view.

To restore this temporary setting, do the following:

8. Click the **Hide/Isolate** icon.

9. Select **Reset Temporary Hide/Isolate** from the pop-up menu.

All is now back to normal – try a few other variations on this feature and observe the results – select Rest Temporary Hide/Isolate before proceeding.

Exercise 6-4:
Basement Floor Plan

Basement floor plan:

Now you will sketch the basement floor plan. Not all residences have basements; they mostly occur in cold climates where the foundation walls are deep to avoid frost heave (once you have dug out for the deep foundation walls along the entire perimeter of the building, it does not take much more effort/resources to create a basement).

Drawing a basement is the same as the other floors; you draw the exterior walls (concrete foundation walls in this case), the interior walls, etc.

Once again, with the residential template that you started with being setup for the typical residential project, it should not surprise you that a typical foundation wall type is already setup and ready to be sketched; both 10" and 12" walls have been pre-defined.

1.   Open your project file ex6-3.rvt and **Save As ex6-4.rvt**.

2.   Switch to the **Basement** *Floor Plan* View.

You should have a blank view (other than the elevation tags).

3.   In *View Properties*, set the **Underlay** *Parameter* to **Second Floor**.

This underlay setting will give you an outline of your building that can be used to sketch the foundation walls.

4.   Using the **Wall** tool, draw the foundation walls using the following information:
     a. *Wall type*: **Basic Wall: Foundation – 12" Concrete**
     b. *Loc Line*: **Finish Face: Exterior**
     c. In *Properties* (before sketching walls); set the *Top Constraint* to: **Up to level: Foundation** (Fig. 6-4.1)
     d. Check **Chain** on the *Options bar*.
     e. Start anywhere, by snapping to one of the exterior corners of the *Underlay*; work in a clockwise direction.

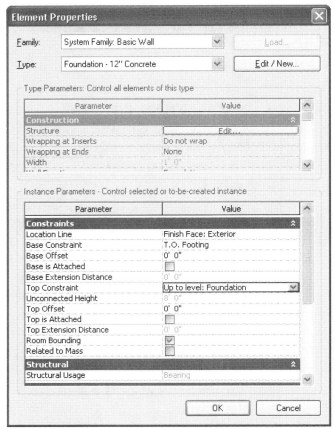

Figure 6-4.1 Basement Floor Plan View Properties

When finished, your plan should look like Figure 6-4.2. The garage area will be slab-on-grade (i.e. no basement under it), so make sure you sketch a foundation wall between the garage and basement as shown.

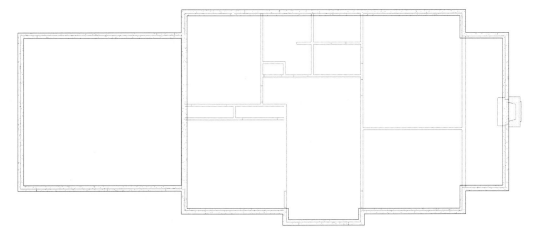

Figure 6-4.2 Basement Floor Plan; with Second Floor Underlay turned on

5. Set the *Underlay* parameter back to **None** in *View Properties*.

Creating a new foundation wall type:

Because the garage area is slab-on-grade, both sides of the foundation wall will have earth against it (the main basement walls act like a retaining wall because they only have earth on one side of the wall). Therefore you will change the three exterior foundation walls, at the garage area, to 8" concrete. You will have to create a new wall type to accommodate this need.

Disclaimer:

All dimensions and sizes shown in this book are solely for the purpose of learning to use the software. Under no circumstance should any information be assumed to be appropriate for a real-world project. A structural engineer should be consulted if you do not have adequate expertise in that area.

---

6. Select the **Wall** tool and then select **Basic Wall: Foundation – 12" Concrete** from the *Type Selector*.

7. Create a new wall type based on the current selection; name it **Basic Wall: Foundation – 8" Concrete**.
   Remember: *Properties → Edit/New → Duplicate*

8. Adjust the Type Parameters of the new wall so the Width is 8".
   FYI: *This is through Properties → Edit/New → Edit Structure.*

9. Select the three exterior garage foundation walls (while holding down the Ctrl key) and then select your new wall in the *Type Selector*.

Your basement plan should look like Figure 6-4.3.

Because the *Loc. Line* was set to Finish Face: Exterior, the exterior side of the foundation wall is stationary when walls of a different width are selected; this is the desired result.

FYI:

You will add footings below the foundation walls in Exercise 10-2.

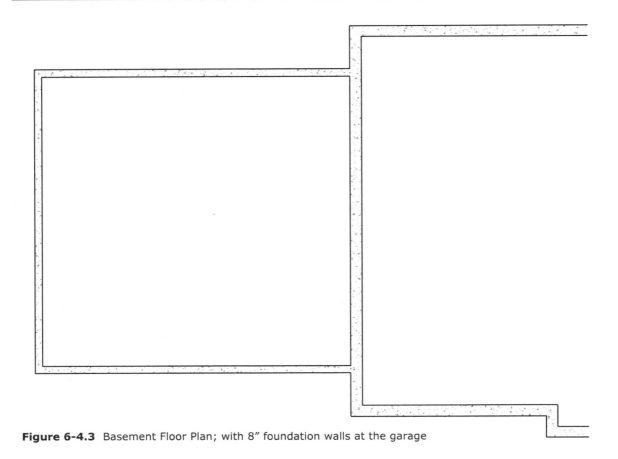

Figure 6-4.3 Basement Floor Plan; with 8" foundation walls at the garage

Adding a few interior walls in the basement:

To make things simple, for this tutorial and structurally, you will copy all the walls from the first floor to the basement. In the process of accomplishing this task you will also study the Filter function.

First you will switch to the first floor and select a large area, using a crossing-window (i.e. picking from right to left) and then use the Filter function to strip the selection down to just the walls.

10. Switch to the first floor and select all the walls using a crossing-window (see Figure 6-4.4).

Everything within or passing through the crossing-window is now selected. Next you will filter out everything except for the walls.

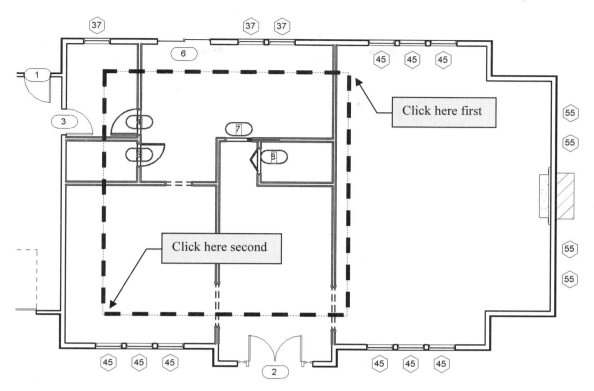

Figure 6-4.4 First Floor Plan; selecting all walls with a crossing-window

11. With everything from the previous step still selected, click the **Filter Selection** icon the *Options Bar*. (Figure 6-4.5)

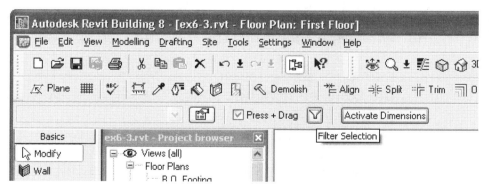

Figure 6-4.5 Filter Selection icon

You are now in the Filter dialog (figure 6-4.6). This lists all the element types (or categories) that are in the current selection set. When you click OK, only the items checked will be part of the current selection set.

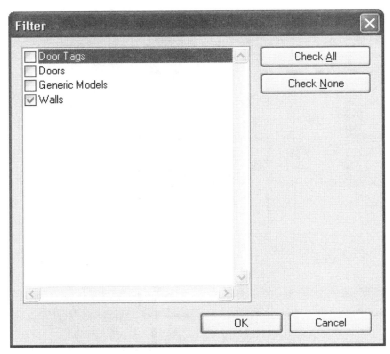

Figure 6-4.6 Filter dialog

12. Uncheck everything except the **Walls** item in the *Filter* dialog. (Figure 6-4.6)

13. Click **OK** to update the current select set.

Now only the walls are selected. Next you will copy them to the Clipboard and then Paste them into the basement floor plan.

This selection technique can be use in many ways. For example, you could select all the walls and change their Top Constraint via the Properties dialog.

14. From the *Edit* pull-down menu select **Copy to Clipboard**.

15. Switch to the basement view.

16. Select **Paste Aligned → Current View**, from the *Edit* pull-down menu.

You may get the following warning (Figure 6-4.7):

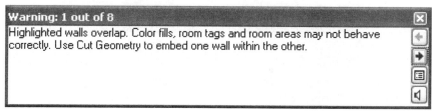

Figure 6-4.7 Wall placement warning

This error is indicating that the wall(s) you just placed overlaps another wall (vertically or horizontally). In this case it may be that the *Top Offset* is not set to 0", but 6"; which causes the wall to extend 6" above the *Top Constraint* (which is the first floor line). So, if the first floor walls start at the first floor line, the basement walls extend up into the first floor walls by 6".

If you get this warning, click the "X" in the upper right corner of the warning window; this will close it. Then select all the interior walls in the basement and make sure the *Top Offset* is set to 0" (via wall properties). You can also check the Base offset for the first floor walls.

When finished, your plan should look like Figure 6-4.8.

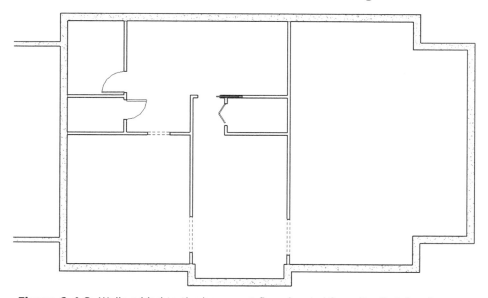

Figure 6-4.8 Walls added to the basement floor (copied from the first floor)

Even though the doors were not selected they were still copied with the walls – just as before. This is fine; to save time you will leave the doors as-is.

17. **Save** your project as **ex6-4.rvt**.

Exercise 6-5:
Stairs

Next you will add stairs to your floor plans. Revit provides a powerful stair tool that allows you to design stairs quickly with various constraints previously specified (e.g., 7" maximum riser).

Pre-defining parameters:

Before you draw the stair it will be helpful to review the options available in the stair family.

1. Open ex6-4.rvt and Save As **ex6-5.rvt**. Open the First floor plan.

2. From the *Project Browser*, expand the Families → Stairs → Stair (i.e., click the plus sign next to these labels).

3. Right-click on the stair type: **Residential – Closed 2 Sides**, and select the **properties...** option from the pop-up menu.

You should now see the options shown in Figure 6-5.1.

Take a couple minutes to see what options are available. You will quickly review a few below.

- Tread Depth: Treads are typically 12" deep (usually code min.) and 1" of that depth overlaps the next tread; so only 11" is visible in plan view . This overlap is called the nosing.
- Riser: This provides Revit with the maximum dimension allowed (by code, or if you want it, less). The actual dimension will depend on the floor-to-floor height.
- Stringer dimensions: These dimensions usually vary per stair depending on the stair width, run and materials, to name a few. A structural engineer would provide this information after designing the stair.
- Cost: Estimating placeholder.

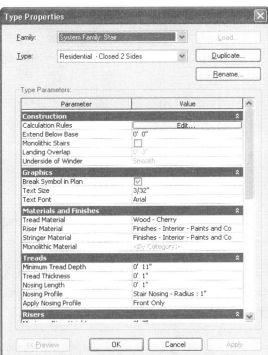

Figure 6-5.1 Stair type properties

Calculate Tread / Riser Size

Although Revit automatically calculates the rise and tread dimensions for you, it is still a good idea to understand what is happening.

The Riser is typically calculated to be as large as building codes will allow. Occasionally a grand stair will have a smaller riser to create a more elegant stair.

Similarly, the Tread is usually designed to be as small as allowable by building codes.

The largest riser and shortest tread creates the steepest stair allowed. This takes up less floor space. (see **Figure 6-5.2**) A stairway that is too steep is uncomfortable and unsafe.

Building codes vary by location; for this exercise you will use 7" (max.) for the risers and 9" (min.) for the treads.

Codes usually require that each tread be the same size, likewise with risers.

Calculate the number and size of the risers:

> Given:
> Risers: **7"** max.   Floor to floor height: **9'-0"** (see elevation view)
>
> Calculate the number of risers:
>   9'-0" divided by 7" (or 109" divided by 7") = 15.429

Seeing as each riser has to be the same size we will have to round off to a whole number. You cannot round down because that will make the riser larger than the allowed maximum (9'-0" / 15 = 7.2"). Therefore you have to round up to 16. Thus: 9'-0" divided by 16 = 6.75

So you need **16** risers that are **6 3/4"** each.

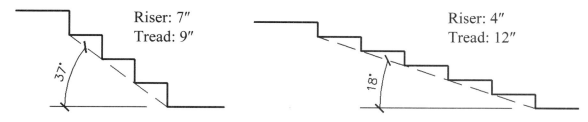

Figure 6-5.2   Stair rise / run comparison

Drawing the stairs in plan:

You will be drawing a u-shaped stair from the first floor to the second and an L-shaped stair from the basement to the first floor. At first, when using Revit to draw stairs, it may be helpful to figure out the number of risers and landings. That information will be helpful when drawing the stair. As you become more familiar with the stair tool you will not need to do those calculations to draw a stair.

To get an idea about what you are going to draw next, take a moment to look at Figure 6-5.3. The image below is at the first floor, and the "UP" label indicates where you would access the stair on the first floor (everything else rises from there, of course, to the second floor).

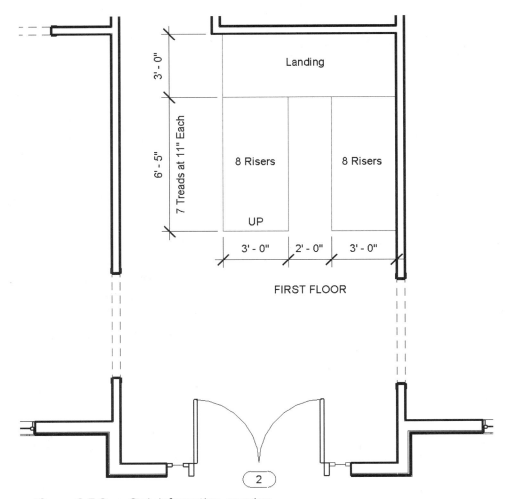

Figure 6-5.3   Stair information overview

The first thing you will do, to prepare to draw the stair from the first floor to the second floor, is to draw a temporary reference line. This line will be used to accurately pick a starting point while using the Stair command (you will be picking a starting point near the middle of the room – next to the "UP" label in Fig. 6-5.3). After the stair is drawn you will erase the line.

You have had plenty of practice drawing lines in previous chapters, so you will be given minimal instruction on how to draw the line.

4.   Using the **Lines** tool from the *Basics* tab, draw the temporary line as shown in Figure 6-5.4.

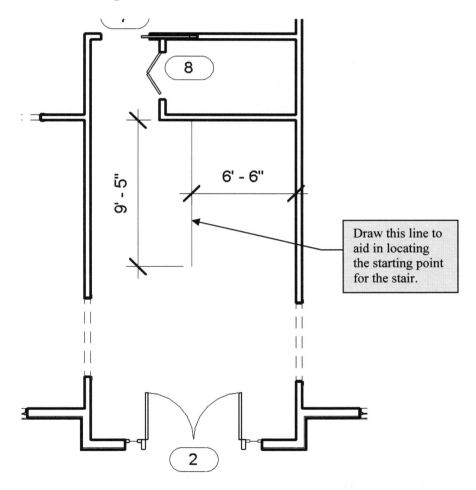

Draw this line to aid in locating the starting point for the stair.

Figure 6-5.4   Draw a temporary reference line

5.   Make sure you are in the **first floor** plan view.

6. **Zoom in** to the front Entry Foyer.

7. Click on the **Modeling** tab in the *Design Bar*.

8. Select the **Stairs** tool (on the *Modeling* tab).

9. Click on the **Stairs Properties** button that appeared in the *Design Bar*. (Figure 6-5.5)

10. Make sure the *Width* is set to **3'-0"** and that *Top Level* is set to **Second Floor**, and then select **OK**. (Figure 6-5.6)

11. Click the endpoint of your temporary line as shown in **Figure 6-5.7**; you are selecting the start point for the first step. Make sure you are snapping to the line for accuracy.

12. Pick the remaining points as shown in Figures 6-5.7, 6-5.8 and 6-5.9.

Figure 6-5.5

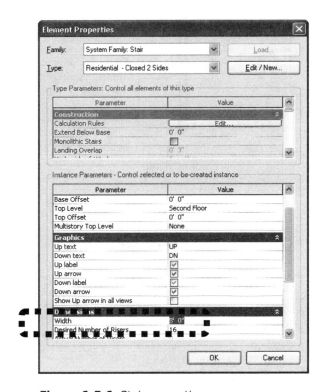

Figure 6-5.6 Stair properties

Notice as you draw the stairs, Revit will display the number of risers drawn and the number of risers remaining to be drawn to reach the next level (or whatever level is selected as *Top Level* in *Properties*).

If you click *Finish Sketch* before drawing all the required risers, Revit will display an error message. You can leave the problem to be resolved later. Revit will make the stair magenta in color until the problem is resolved.

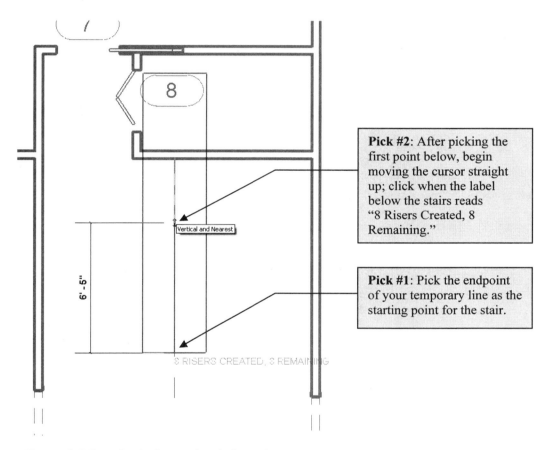

Vertical and Nearest

Pick #2: After picking the first point below, begin moving the cursor straight up; click when the label below the stairs reads "8 Risers Created, 8 Remaining."

Pick #1: Pick the endpoint of your temporary line as the starting point for the stair.

8 RISERS CREATED, 8 REMAINING

Figure 6-5.7   Beginning to sketch the stair

As soon as you click *Finish Sketch*, Revit will automatically add the landing; creating it between to two stair runs sketched. The width of the landing will match the stair width, which is 3'-0". This is true for any stair configuration using the above method. You will see this momentarily when you sketch an L-shaped stair in the basement.

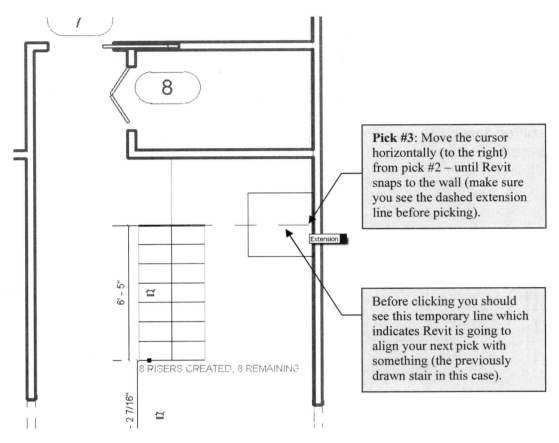

Pick #3: Move the cursor horizontally (to the right) from pick #2 – until Revit snaps to the wall (make sure you see the dashed extension line before picking).

Before clicking you should see this temporary line which indicates Revit is going to align your next pick with something (the previously drawn stair in this case).

Figure 6-5.8   Continuing to sketch the stair

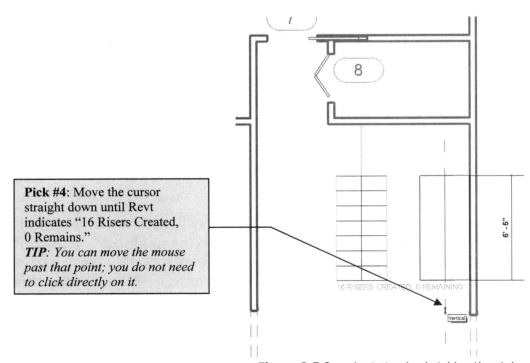

Pick #4: Move the cursor straight down until Revt indicates "16 Risers Created, 0 Remains."
TIP: You can move the mouse past that point; you do not need to click directly on it.

Figure 6-5.9   Last step in sketching the stair

13. Click **Finish Sketch**.
   (Figure 6-5.5)

14. Erase the Temporary line.

Your stair should now look like Figure 6-5.10! Notice that Revit adds a break-line at the *Cut Plain* and everything above that is dashed in.

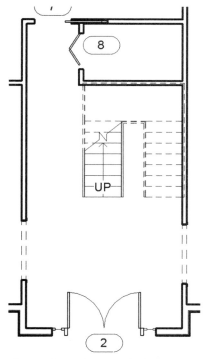

Figure 6-5.10   Finished stair

Drawing the basement stairs:

Now you will draw the L-shaped stairs from the basement to the first floor.

To get an idea about what you are going to draw next, take a moment to look at Figure 6-5.11. The image below is at the basement level, and the "UP" label indicates where you would access the stair on the basement level (everything else rises from there, of course, to the first floor).

15. Save your project as **ex6-5.rvt**.

16. Switch to the **Basement** floor plan view.

17. Select the **Stair** tool.

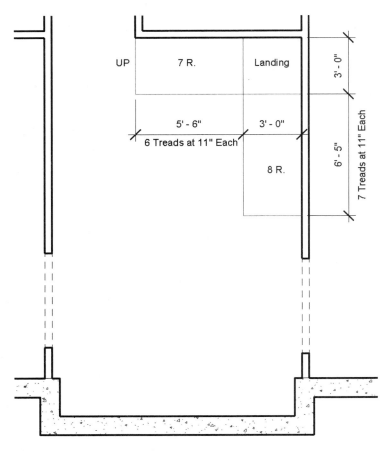

Figure 6-5.11   Stair information overview

18. Select the **Stair Properties...** button on the *Design Bar*. (Figure 6-5.5)

19. In the *Properties* dialog:
    a. Change the *Type Selector* to **Residential – Open 2 Sides**.
    b. Make sure the **Top Level** is set to **First Floor**; the *Stair* tool defaults to the next *Level* up – which is "foundation" level and not "first floor" in your case.

The stair type "Residential – Closed 2 Sides" (which should have been the default for the first floor stair) is setup to add a guardrail to both sides of the stair; you will see that in the 3D image at the end of this exercise.

Therefore, the stair you are drawing in the basement will not have a guardrail on either side of the stair.

20. Using the same techniques you employed to sketch the first floor stair, draw the basement stair per the information given in Figure 6-5.11. *TIP: While sketching the stairs, you can select a window (picking from left to right) around a run of stairs and move it. Also, don't forget to draw a temporary line(s) if you think it will help).*

Your stair should look like Figure 6-5.12. If not, select one of the riser/tread lines and press the Delete key on the keyboard to erase the stair and try again. It can get a little frustrating when you are not familiar with the ins-and-outs of what Revit is doing or what information it needs to proceed. Take your time, take a break if necessary.

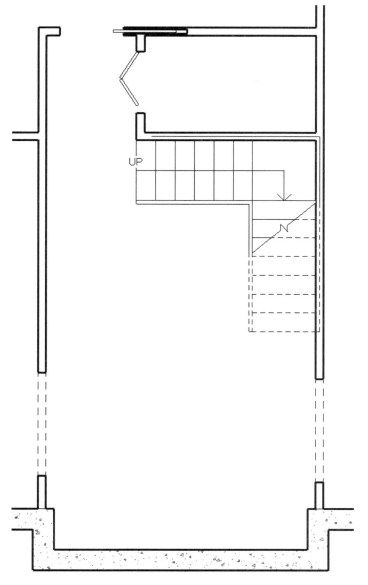

Next you wall draw three walls next to the stair. This will close off the two open sides and the bottom of the stair.

FYI (again): This tutorial is not meant to be a practical design as an L-shaped stair with walls would make it more difficult to move large items into the basement (e.g. washer, dryer, furniture, etc.).

Figure 6-5.12   Finished basement stair

21. Using the **Wall** tool, draw the three interior walls as shown in **Figure 6-5.13**.
   - Use the standard interior wall you used for the other floors.
     (***TIP**: You can select one and see what type it is on the Options Bar.*)
   - Click the *Chain* option.
   - Set *Loc. Line* to one of the wall faces.
   - *Top Constraint* set to: First Floor.
   - Click the inside points of the stringer, so the stringer gets buried in the wall.
   - Draw the wall under the stair, at the fourth riser line.

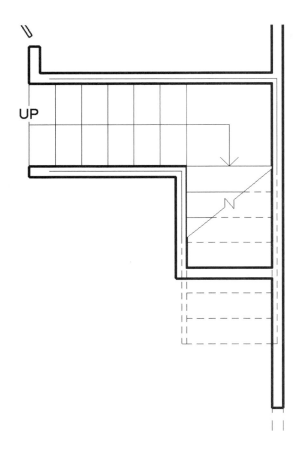

Figure 6-5.13   Three walls added at stairs

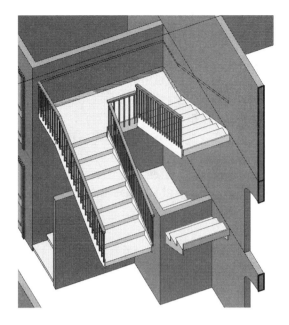

Figure 6-5.14   3D view of stair

A cutaway 3D view of your model looks something like this (you will learn to do cutaway views later).

You should notice two things about this 3D view:

First: the basement wall you just added extends through the basement stair. You will fix that in the building sections lesson.

Second: no floors have been drawn yet. You will draw floor systems in the next chapter.

22. **Save** your project.
    FYI: Normally you would not rename your project file – you do it here in the book for grading and restore options, should thing get really messed up.

Be sure to check out the sample stair project file on Revit's online content library (revit.autodesk.com/library/html). You can download this file and see examples of several different stair types side-by-side (Figure 6-5.15). You can select one and view its properties to see how it is done. You can also Copy/Paste one into your project, select your stair, and then select the newly imported type(s) from the Type Selector. The partial view of the sample file, shown below, has open riser, single stringer, no stringer, spiral, etc.

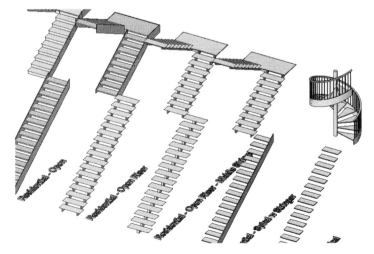

Figure 6-5.15   Partial view of sample stair file
from Revit's website

Modify the buildings floor-to-floor height:

You will not need to do this for this tutorial, but you can modify the buildings floor-to-floor height (or Level). The reasons for doing this vary. Some examples might be to make the building shorter or taller to accommodate ductwork in the ceilings or the depth of the floor structure (the longer the span the deeper the structure). The default floor-to-floor height in the template file you started from is 9'-0", which is fine for this project. However, should you want to change this for another project, here's how its done:

- Open any exterior elevation from the *Project Browser*.

- Change the floor-to –floor height to be 12'-0" for each level (in this example to the right).

- Select the floor elevation symbol, and then select the text displaying the elevation. You should now be able to type in a new number. Press Enter to see the changes. Notice the windows move because the sill height has not been changed. (Figure A)

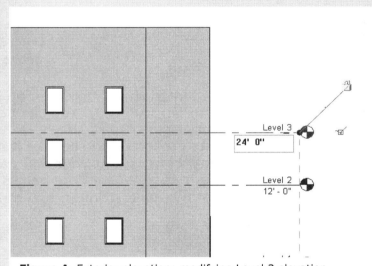

Figure A Exterior elevation: modifying Level 3 elevation

The floor-to-floor height is a good thing to figure out (when possible) before drawing the stairs as it affects the number of risers and treads required; not to mention floor area.

Exercise 6-6:
Dimensions

Next you will add dimensions to your floor plans. Revit provides a powerful dimension tool that allows you to quickly dimension drawings. This process is much easier than other CAD programs. Revit has a dimension style all setup and ready to use; which works for most designers. Furthermore, the size of the text and ticks are tied to the View Scale feature to make sure the dimensions always print at the correct size. For example, a dimensions' text and ticks would get twice as big just by changing the View Scale from 1/4" = 1'-0" to 1/8" = 1'-0". You will try this in a moment.

Industry standard dimensioning conventions were covered in the previous Lesson, you may want to take a minute and review that information (see page 5-23).

Drawing dimensions in Revit involves the following steps:

- Select the Dimensions tool from the Basics tab

- Options Bar:

   o Select one of the predefined styles from the Type Selector.
   o Select the kind of dimension you want (e.g. Linear, Angular, Radial, Arch Length).
   o Tell Revit what part of the wall you want to dimension to (e.g. Wall centerlines, Wall Face, Center of Core, Face of Core); this can change for each side of a dimensions line.

- Select two walls to dimension between.

- Select a location for the dimension line.

Adding dimensions to the First Floor Plan:

1.   Open ex6-5.rvt and **Save-As ex6-6.rvt**.

2.   Switch to the first floor plan view.

3. Zoom in on the north wall of the garage.

4. Select the **Dimension** tool from the *Basics* tab.

Notice the Options Bar changes to give you control over the various options mentioned above (Figure 6-6.1).

Figure 6-6.1   Option Bar: dimension tool active

5. Set the *Type Selector* to **Linear Dimension Style: Linear – 3/32" Arial**.

6. Click the **Properties** button on the *Options Bar*.

7. Click **Edit/New...**

Notice the various properties you can adjust for the selected dimension type (Figure 6-6.2).

Creating a new, custom, dimension type is the same as for walls, doors, etc. – just click the Duplicate button, provide a name and then adjust the properties.

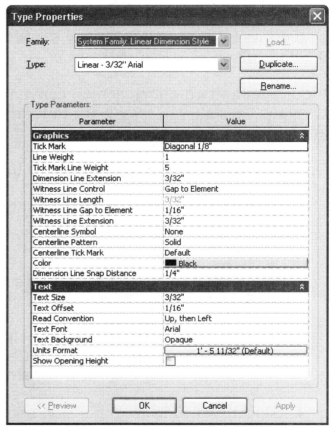

Figure 6-6.2   Properties: for selected dimension type

All the dimensions you see here are multiplied by the *View Scale*. For example, the 3/32" text is multiplied by 48 when the *View Scale* is set to 1/4" = 1'-0"; thus, the text is 4 ½" tall in 1/4" drawings. When the 1/4" view is placed on a *Sheet*, Revit automatically scales the view down 1/48 (at this point the text is 3/32" again). This makes the view actually a 1/4" = 1'-0" on the Sheet (which is actual sheet size – say 22" x 34"). Finally, all this is changed automatically whenever the *View Scale* is changed (dimension and text size as well as the *View* size on sheets)! *FYI: You will learn about Sheets later in the book.*

8. Click **OK** twice to close the dialog boxes.

9. On the *Options Bar* select the **Linear** icon.
   a. *(Just to the right of the properties button)*

Linear allows you to draw horizontal, vertical and angled dimensions.

10. Next to *Prefer*, select **Wall Faces** on the *Options Bar*.

11. Draw your first dimension by clicking the three points as identified in **Figure 6-6.3**.

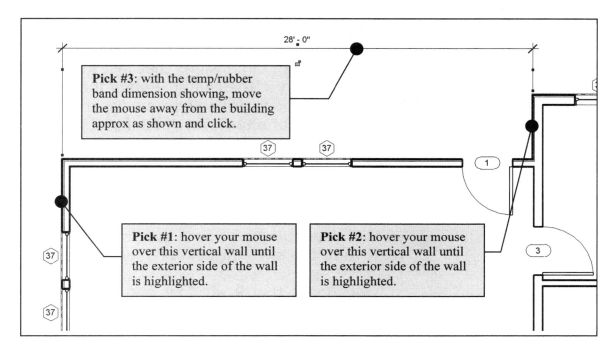

Figure 6-6.3   First Floor plan: first dimension added

FYI: In the figure above (6-6.3), you can select points 1 and 2 in the opposite order as well.

12. Dimension the rest of the north exterior wall as shown in Figure 6-6.4 (see the next page for dimension comments).

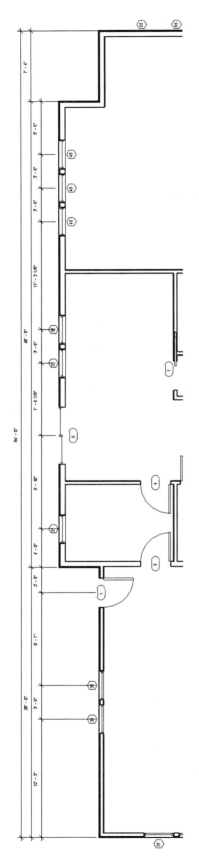

Once you have drawn one dimension, as in step 11, the remaining dimensions in that "string" should align with the first. Revit helps you make the third pick (as in Figure 6-6.3) by automatically snapping to the previous dimension when you move your cursor to it.

Yet another feature Revit offers that other programs don't, is the ability to automatically snap to the center of a door or window opening. In other programs you might have to draw a line from one door jamb to the other, use the dimension command to snap to the mid-point of the line and then delete it. Then you are left with dimensions that just look like they dimension to the center. Remember, Revit's dimensions are able to modify the drawing (kind of a two-way road if you will), whereas other CAD programs are mostly a one-way road.

In addition to snapping to the center of doors and windows, Revit will also snap to the jambs; depending on where your mouse is relative to the element being dimensioned.

You will notice that Revit does not care if a dimensions witeness line overlaps a door or window tag. The tags need to be selected and moved manually (individually or in groups).

Next you will dimension the west side of the building.

Figure 6-6.4
First Floor: rotated 90 degrees to increase image size in book

13. Dimension the west exterior walls and openings as shown in **Figure 6-6.5**.

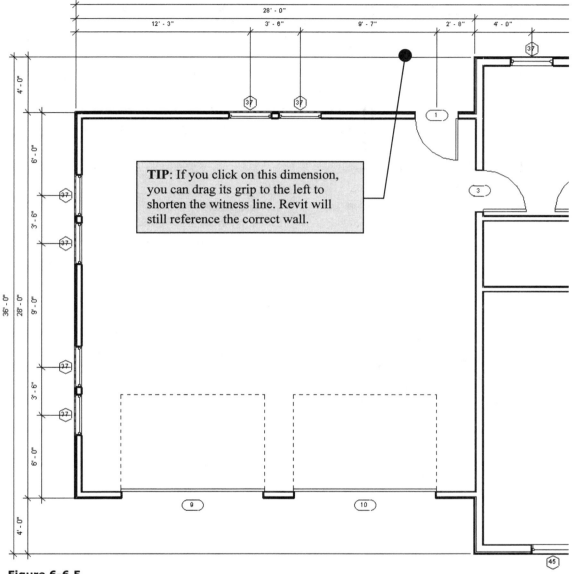

TIP: If you click on this dimension, you can drag its grip to the left to shorten the witness line. Revit will still reference the correct wall.

Figure 6-6.5
First Floor: dimensions added to the west wall

The last two images in the exercise give you an idea of what each dimension type (that was loaded with the residential template) looks like. Similar to other elements in Revit, you can select a dimension and verify what type it is by looking at the *Type Selector* on the *Options Bar*; also, you can change the type by selecting the desired type from the *Type Selector*.

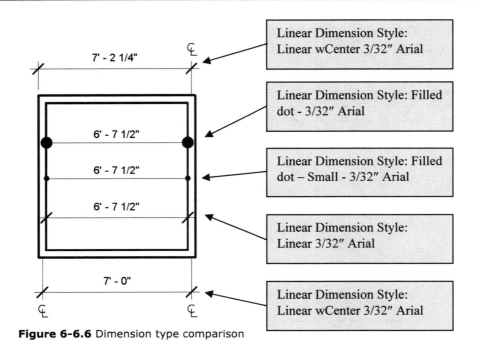

Figure 6-6.6 Dimension type comparison

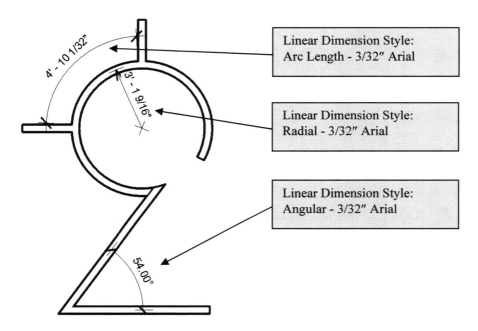

Figure 6-6.7 Dimension type comparison

This should be enough for you to understand how to dimension your plans (or anything else). If you want more practice you can dimension the rest of the first floor plan, and then the other two floors (your instructor may require this).

14. **Save** your Project as **ex6-6.rvt**.

Self-Exam:

The following questions can be used as a way to check your knowledge of this lesson. The answers can be found at the bottom of this page.

1. The default settings for the floor plan view shows the walls for the floor below. (T/F)

2. The size of Dimension text and arrows are set by the View Scale feature. (T/F)

3. You should start drawing your floor plan generally centered on the default elevation tags in a new project. (T/F)

4. You can use the Align tool to align one wall with. (T/F)

5. Where do you change the maximum riser height? _____

Review Questions:

The following questions may be assigned by your instructor as a way to assess your knowledge of this section. Your instructor has the answers to the review questions.

1. It is not possible to copy/paste objects from one floor to another and have them line-up (with the original objects). (T/F)

2. Revit's dimension tool can snap to the middle of a door opening without the need to draw a temporary line to snap to. (T/F)

3. Each Revit view is saved as a separate file on your hard drive. (T/F)

4. You select the part of the wall to be deleted when using the Trim tool. (T/F)

5. You cannot "finish sketch" if you have not drawn all the tread/riders that are required between the two specified levels. (T/F)

6. What parameter should be set to none, in the view properties dialog, if you

   do not want to see the walls from the floor below? _____

7. The _____ _____ controls the size of dimensions and text, plus the scale that view is placed on a sheet.

8. You can use the _____ tool to quickly select a certain type of object from a large group of selected objects.

9. The number of _____ remaining is displayed while sketching a stair.

Self-Exam Answers:
1 - T, **2** - T, **3** - T, **4** - T, **5** - Properties

Notes:

Lesson 7
Residence: ROOF::

This lesson will look at some of the various options and tools for designing a roof for your building. You will also add skylights.

Exercise 7-1:
Roof Design options (Style, pitch and overhang)

In this lesson you will look at the various ways to use the *Roof* tool to draw the more common roof forms used in architecture today.

Start a new Revit Project:

You will start a new project for this lesson so you can quickly compare the results of using the *Roof* tool.

1. Start a new project using the **default.rte** template.

2. Switch to the **North** elevation view and rename the *Level* named Level 2 to **T.O. Masonry**. This will be the reference point for your roof. *TIP: Just click on the level tags text to rename.*

3. Switch to the **Level 1** floor plan view.

Draw walls to setup for using the roof tool:

4. Set the Level 1 *"Detail Level"* to **medium**, so the material hatching is visible within the walls. *TIP: Right-click on Level 1 in project browser & select Properties or use the View Ctrl Bar at the bottom.*

5. Using the **Wall** tool and wall type set to **"Exterior - Brick on Mtl. Stud,"** draw a **40'-0" x 20'-0"** building. (Figure 7-1.1) *FYI: The default Wall height is ok (it should be 20'-0").*

Be sure to draw the
building within the
elevation tags.

*TIP: You can draw the
building in one step if you
use the rectangle option on
the Options Bar (while using
the Wall tool).*

Elevations tags – (4) are shown which
correspond to the (4) elevation views
listed in the Project Browser

Figure 7-1.1   Building and Elevation tags

You will copy the building so that you have 4 total. You will draw a
different type of roof on each one.

6.   Drag a window around the walls to select them. Then use the
     Array command to setup four buildings **35'-0" O.C.**
     (Figure 7-1.2)

*TIP: Zoom in and make
sure the brick is on the
exterior side of the wall,
if not you can select
each wall and click its
flip icon.*

*ARRAY TIP: Select the
first building, select
Array, and then, just
like the Copy command,
define a copy 35' to the
right, and then enter
the number of copies.*

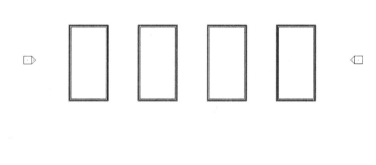

Figure 7-1.2   4 buildings

7.   Select all of the buildings and click **Ungroup** from the *Options
     Bar*.

Hip roof:

The various roof forms are largely defined by the "Defines Slope" setting. This is displayed in the *Options Bar* while the *Roof* tool is active. When a wall is selected, while the "Defines slope" option is selected, the roof above that portion of wall slopes. You will see this more clearly in the examples below.

8. Switch to the **T.O. Masonry** view.

9. Select the **Roof** tool, and then **Roof by Footprint** from the pop-up menu that appears after selecting the Roof tool.

10. Set the overhang to **2'-0"** and make sure "*Defines slope*" is selected (checked) in the *Options Bar*.

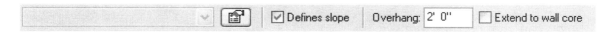

11. Now select the four walls of the building at the far left.
    TIP: Make sure you select towards to exterior side of the wall; notice the review line before clicking.

12. Click **Finish Roof**.

13. Click **Yes** to attach the Roof to the walls.

14. Switch to the **South** elevation view. (Figure 7-1.3)

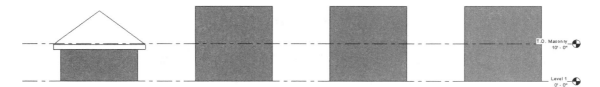

Figure 7-1.3   South elevation – hip roof

You will notice that the default wall height is much higher than what we ultimately want. However, when the roof is drawn at the correct elevation and you attach the walls to the roof, the walls automatically adjust to stop under the roof object.

15. Switch to the **3D** view using the icon on the *View* toolbar. (Figure 7-1.4)

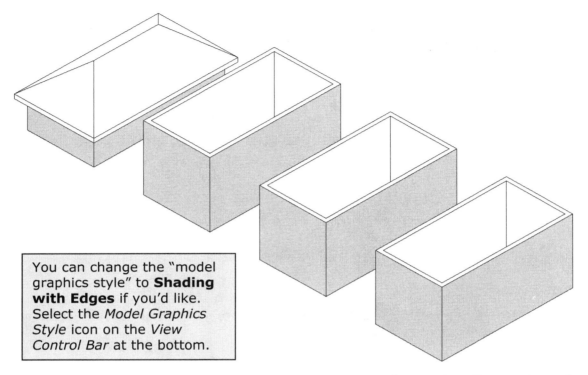

> You can change the "model graphics style" to **Shading with Edges** if you'd like. Select the *Model Graphics Style* icon on the *View Control Bar* at the bottom.

Figure 7-1.4   3D view – hip roof

Gable roof:

16. Switch back to the **T.O. Masonry** view.

17. Select the **Roof** tool, and then **Roof by Footprint**.

18. Set the overhang to **2'-0"** and make sure "*Defines slope*" is selected (checked) in the *Options Bar*.

19. Only select the two long (40'-0") walls.

20. Uncheck the "*Defines slope*" option.

21. Select the remaining two walls. (Figure 7-1.5)

22. Pick **Finish Roof**.

23. Select **Yes** to attach the walls to the roof.

24. Switch to the **South** elevation view.
    (Figure 7-1.6)

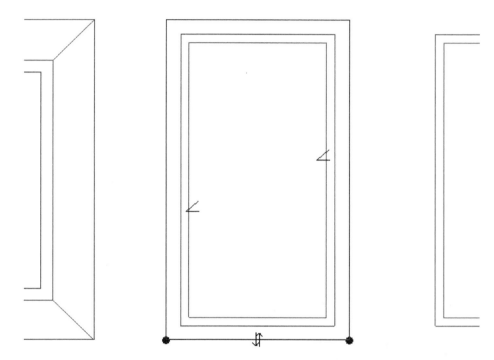

Figure 7-1.5 Gable – plan view

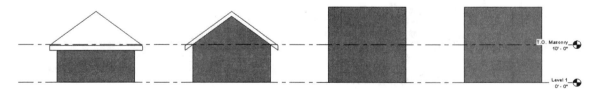

Figure 7-1.6 South elevation – gable roof

25. Switch to the **3D** view (Figure 7-1.7).

Notice the wall extends up to conform to the underside of the roof on the gable ends.

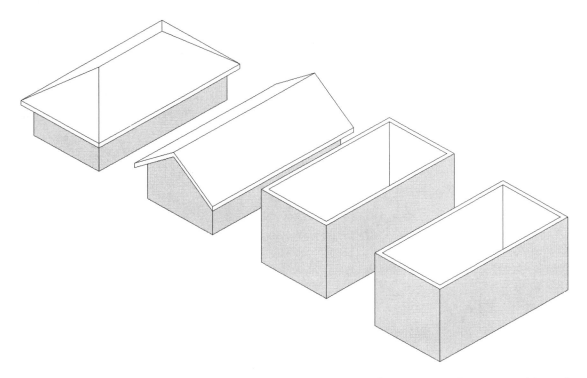

Figure 7-1.7   3D view – gable roof

Shed roof:

26. Switch back to the **T.O. Masonry** view.

27. Select the **Roof** tool, and then **Roof by Footprint**.

28. Check "*Defines slope*."

29. Click on the **Properties** button on the *Options Bar*.

30. Set the roof pitch to 3/12 (Figure 7-1.8); click **OK**.
    FYI: *The "Defines slope" option must be selected to change the pitch.*

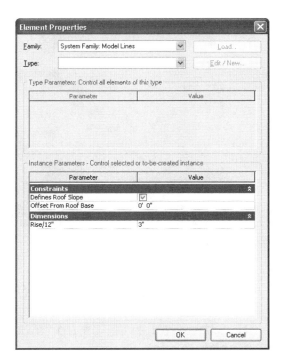

Figure 7-1.8   Properties for roof tool

31. Set the overhang to **2'-0"** and make sure "*Defines slope*" is selected (checked) in the *Options Bar*.

32. Select the east wall (40'-0" wall, right).

33. **Uncheck** "*Defines slope*" in the *Options Bar*.

34. Select the remaining three walls. (Figure 7-1.9)

35. Pick **Finish Roof**.

36. Select **Yes** to attach the walls to the roof.

FYI: *You can also change the pitch of the roof by changing the Pitch Control text (see Figure 7-1.9); just select the text and type a new number.*

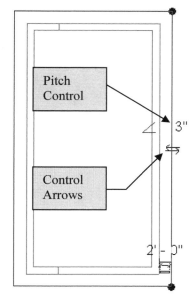

Figure 7-1.9   Selected walls

TIP: You can use the **Control Arrows** (while the roof line is still selected) to flip the orientation of the roof overhang if you accidentally selected the wrong side of the wall (and the overhang is on the inside of the building). (Figure 7-34)

---

37. Switch to the **South** elevation view. (Figure 7-1.10)

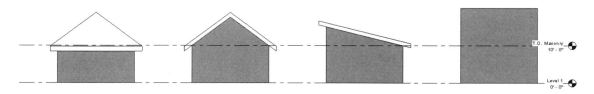

Figure 7-1.10   South elevation – shed roof

38. Switch to the **3D** view (Figure 7-1.11)

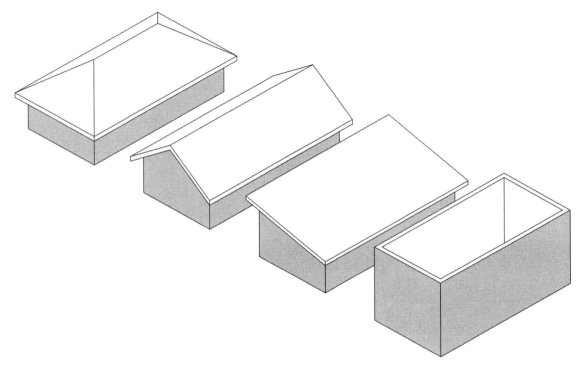

Figure 7-1.11   Default 3D view – shed roof

Once the roof is drawn, you can easily change the roof's overhang. You will try this on the shed roof. You will also make the roof slope in the opposite direction.

39. In **T.O. Masonry** view, Select **Modify** from the *Design Bar*, and then select the shed roof.

40. Click **Edit** from the *Options Bar*.

41. Click on the east (right) roofline to select it.

42. **Uncheck** "*defines slope*" from the *Options Bar*.

43. Now select the west roofline and check "*defines slope*."

If you were to select Finish Roof now, the shed roof would be sloping in the opposite direction. But, before you do that, you will adjust the roof overhang at the high side.

44. Click on the east roofline again, to select it.

45. Change the overhang to **6'-0"** in the *Options Bar*.

Changing the overhang only affects the selected roofline.

46. Select **Finish Roof**.

47. Switch to the South view to see the change. (Figure 7-1.12)

That concludes the shed roof example.

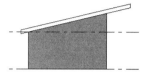

Figure 7-1.12
South elevation –
shed roof (revised)

Flat roof:

48. Switch back to the **T.O. Masonry** view.

49. Select the **Roof** tool & then **Roof by Footprint**.

50. Set the overhang to **2'-0"** and make sure "*Defines slope*" is not selected (un-checked) in the *Options Bar*.

51. Select all four walls.

52. Pick **Finish Roof**.

53. Select **Yes** to attach the walls to the roof.

Figure 7-1.13   South elevation – flat roof

54. Switch to the **South** elevation view. (Figure 7-1.13)

55. Also, take a look the **3D** view. (Figure 7-1.14)

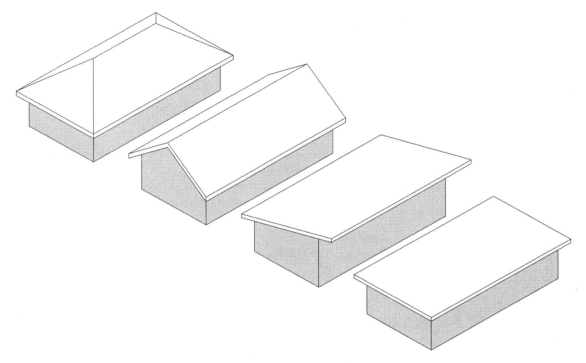

Figure 7-1.14   Default 3D view – flat roof

56. Save your project as **ex7-1.rvt**.

Want more?

Revit has additional tools and techniques available for creating more complex roof forms. However, that is beyond the scope of this book. If you want to learn more about roofs, or anything else, take a look at one of the following resources:

- Revit **Tutorials** from the *Help* pull-down menu
- Revit **Web Site** (www.autodesk.com)
- Revit **Newsgroup** (potential answers to specific questions)
  (www.augi.com; www.revitcity.com; www.autodesk.com)

Exercise 7-2:
Gable roof

The template file you started with already has a Roof Plan view created. You can think of this as the top plate height for your wall. This Roof Plan view creates a working plane for the Roof tool. After creating a high roof (over the two story areas), you will create a "Low Roof" view and design roofs over the one story areas (garage and portion of the family room). Finally, you will create a shed roof for the front porch.

Create Hip Roof:

1. Open ex6-6.rvt and **Save As ex7-2.rvt**.

2. Switch to the **Roof Plan** view.

Your image should look something like Figure 7-2.1. The walls you can see in the Roof Plan view were creating in Lesson 5; the specified height extends into the Roof Plan's View Range. Whereas the two walls you can't see were created on the second floor, in Lesson 6, and the specified height does not extend up into the Roof Plan's View Range. This does not really matter because, as you will see in a moment, you have enough information visible to create the footprint of the roof; also, you will Attach the walls to the roof (just like the previous exercise), which causes them to conform to the underside of the roof,

Because the walls are not connected, neither will the overhang lines you will be sketching. Before you can "finish" a roof sketch, you must have the entire perimeter of the roof drawn with line endpoints connected. Revit allows you to sketch additional lines and use tools like Trim to complete the perimeter of the roof sketch.

3. Select the **Roof** tool from the *Basics* tab.

4. Select **Roof by Footprint** from the pop-up menu.

5. Make sure **Pick Walls** is selected on the *Design Bar*.

Figure 7-2.1 Roof Plan; initial view

6. Click the **Properties** button on the *Options* Bar.

7. Change the roof pitch (Rise/12") to **8"**.

8. Click **OK**.

9. Set the *Overhang* to **1'-6"** on the *Options* Bar.

You only need to select four walls to define the extents of the gable roof you are about to create. However, the perimeter will not be closed after selecting the four walls.

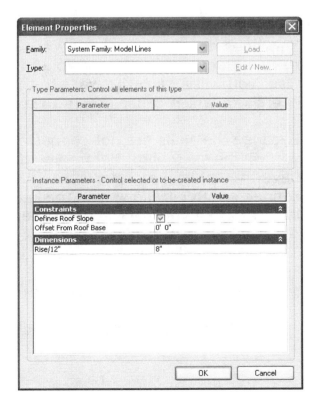

Figure 7-2.2 Roof Properties

10. With "Defines Slope" selected on the *Options Bar*, select the two walls as shown in Figure 7-2.3, make sure the overhang lines are on the exterior side of the building *TIP: Use the flip control icon while the line is still selected if your overhang is on the wrong side (i.e. interior).*

11. With "Defines Slope" NOT selected on the *Options Bar*, Select the two walls as shown in Figure 7-2.3, make sure the overhang lines are on the exterior side of the building.

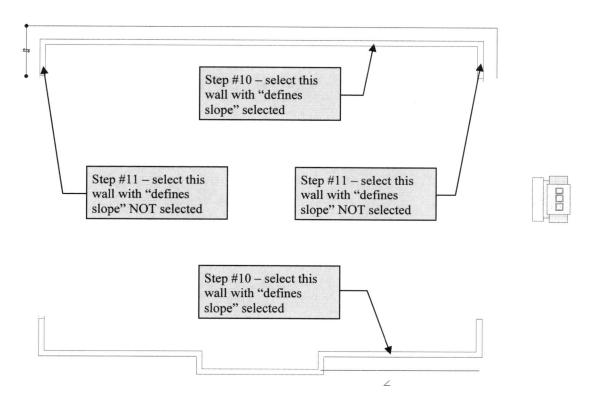

Figure 7-2.3 Roof Plan; four walls select for sketching roof footprint

Note that the four walls, just selected, can be selected in any order. Of course it would be easier to select all the sloped edges first and then all the non-sloping edges second. That way you don't have to keep toggling the "defines slope" option on and off.

Next you will use the *Trim* tool to close the perimeter of the roof footprint. (Many of your 2D drafting skills learned early in this book can be applied to this type of sketch mode.)

12. Select **Modify** from the *Design Bar*.

13. Select the **Trim** icon, and then select the roof lines that are not yet connected to make them connected.

Figure 7-2.4 Roof Plan; entire roof perimeter defined

Now that the roof footprint is complete you can finish the roof.

14. Select **Finish Roof** from the *Design Bar*.

15. Click **Yes**, to attach walls. (Figure 7-2.5)

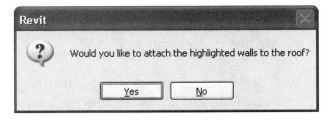

Figure 7-2.5 Roof Plan; attachment warning

16. Switch to your **3D view**.

Your Project should look like Figure 7-2.6. Notice that the walls that were visible in the Roof Plan view are now attached to the roof. The east and west walls that have not been attach can be attached manually.

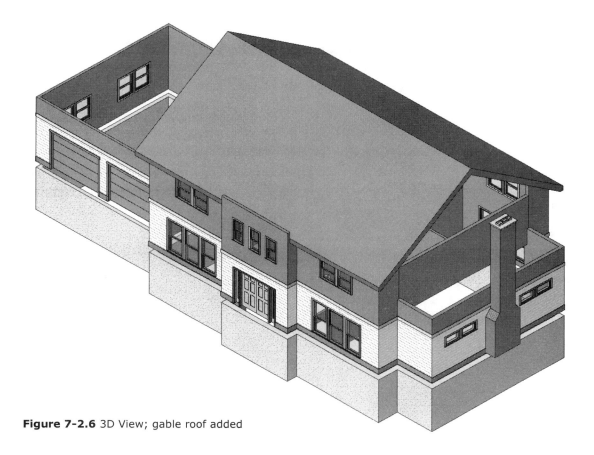

Figure 7-2.6 3D View; gable roof added

17. While in the 3D view, select the east wall that has not been attached to the roof yet.

18. Select the **Attach** button from the *Options Bar* and then select the roof.

19. Repeat the previous two steps with the west wall.

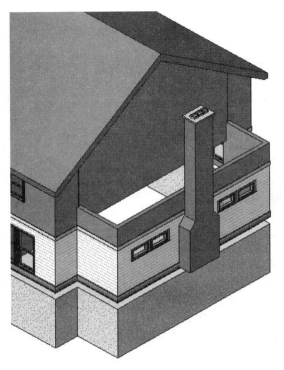

Figure 7-2.7 3D View; wall attached to roof

Main Entry Gable:
Next you will draw the gable roof over the main entry.

You can create a roof on a lower level if the objects, such as walls, are not visible in the Roof Plan view. You simply set the Base Level parameter while sketching the roof. You will try this next.

20. Switch to the **Second Floor** view.

21. Select the **Roof** tool, and then **Roof by Footprint**.

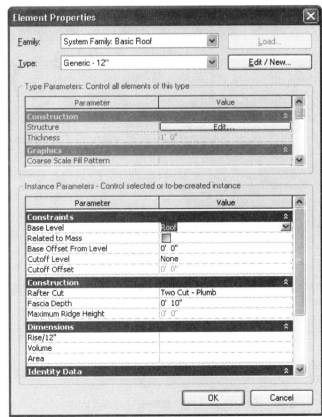

22. Select **Roof Properties** on the *Design Bar*.

23. Set the *Base Level* to **Roof** (see Fig 7-2.8).

Notice the other features of the roof that you can control. (Figure 7-2.8)

24. Select **OK**.

Figure 7-2.8 Element Properties; Roof Tool

Next you will pick the three walls at the main entry to start defining the roof footprint.

25. Select the three walls shown in Figure 7-2.9 per the following:
    a. The two vertical walls "define slope."
    b. The horizontal wall does not "define slope."

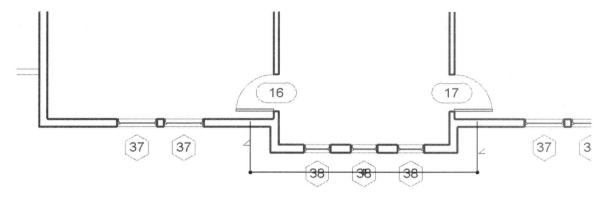

Figure 7-2.9 Second Floor plan view; three walls select to define roof footprint

Now you must sketch a horizontal line at the top edge to create a closed perimeter.

26. Select the **Lines** command on the *Design Bar*. (Figure 7-2.10)

27. Sketch a horizontal line as shown in **Figure 7-2.11**; the exact location does not matter.

28. Use the **Trim** tool to extend the two vertical lines up to the new horizontal line. (Figure 7-2.12)

29. Select **Finish Roof**.

NOTICE: *You cannot see the roof on the second floor. This is because you drew the roof with the Base Level set to "roof"; which is just like drawing in the roof plan view.*

Figure 7-2.10
Design Bar

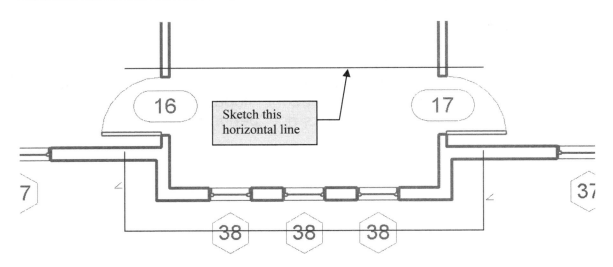

Figure 7-2.11 Second Floor plan view; fourth line sketch (to aid in closing footprint)

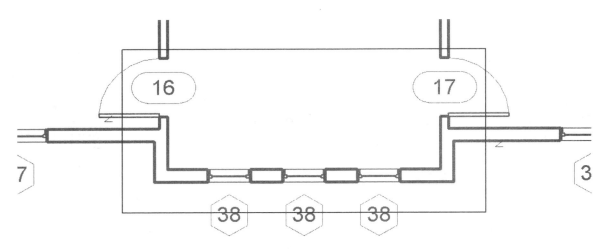

Figure 7-2.12 Second Floor plan view; roof footprint closed

The main entry gable is almost done. You will see what still needs to be done when you switch to the 3D view.

    30. Switch to the **3D** view.

Notice that the two roofs do not intersect correctly (Figure 7-2.13). Revit provides a tool, Join/Unjoin Roof, which will fix this problem.

    31. From the *Tools* pull-down menu, select **Join/Unjoin Roof**.

32. Select the back edge of the new roof. (Figure 7-2.13)

33. Next, you are prompted to select the roof surface you want to join to (see status bar), move your cursor near the edge of the main roof (until that face of the roof highlights) and then click.

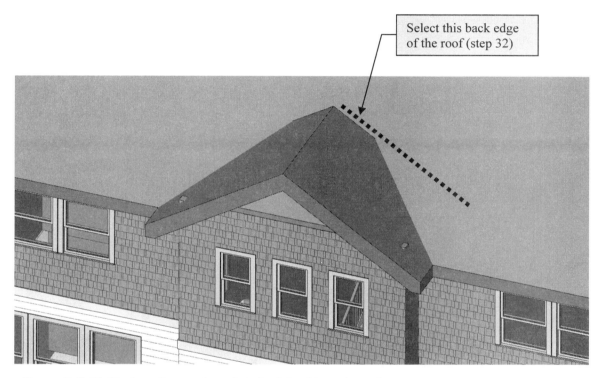

Select this back edge of the roof (step 32)

Figure 7-2.13 3D view; New roof needs to be Joined to main roof

Your roof should now be nicely joined together. Next you need to fix the walls below the roof by attaching them to the new roof.

34. Attach the three walls below the roof using techniques previously covered.

When finished, your 3D view should look like Figure 7-2.14.

Fascia alignment problems?

If, for some reason, your two roof elements do not align at the fascia, it may be that one of the parameters in Properties is different for one of the roof elements (e.g. base offset, rafter cut, fascia depth). Select one roof element and click properties; review the various settings. Now do the same for the other roof and note any differences. Change the settings to match.

Figure 7-2.14 3D view; Roof joined and walls attached

35. Save you project as **Ex7-2.rvt**.

TIP: You have probably figured this out by now: the last view open when you close will be the view that is opened the next time the project is opened.

Exercise 7-3:
Low roof elements

In this exercise you will add the lower roof elements, which will involve adding a new level for reference.

Create a new level:

1. Open ex7-2.rvt and **Save As ex7-3.rvt**.

2. Open the **South elevation view**.

3. Click on the **Level** tool from the *Design Bar's Basic* tab.

4. Draw a Level symbol at the top of the low exterior walls, at elevation 12'-0". Draw the symbol so both ends align with the other symbols below/above it (Figure 7-3.1).

Figure 7-3.1 South Elevation; new level added

Notice that Revit automatically names the level and creates a floor plan and reflected ceiling plan view with the same name (Figure 7-3.2). Next, you will rename the Level.

5. Press **Esc** or select **Modify** from the *Design Bar*.

6. Now select the *Level* symbol you just created.

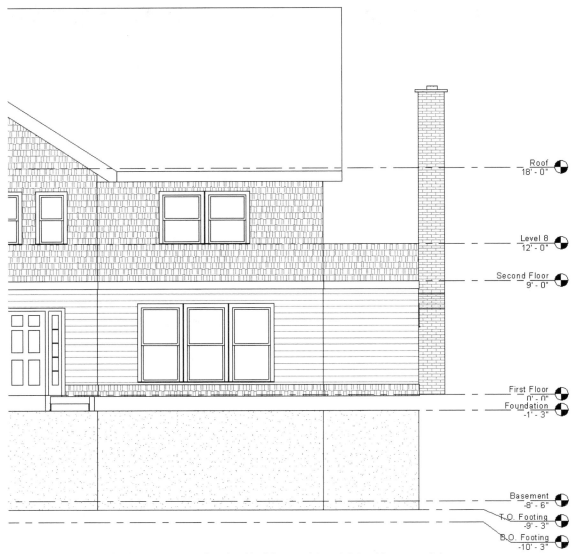

Figure 7-3.2 South Elevation; new level added (named Level 8 in this example)

7. With the level symbol selected, click on the text to rename the level label.

8. Change the label to **Low Roof**. (Figure 7-3.3)

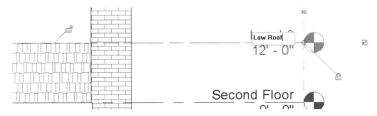

Figure 7-3.3 South Elevation; renaming a Level view

9. Click **OK** when prompted to rename corresponding views. (Figure 7-3.4)

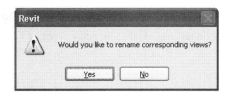

Figure 7-3.4 Rename prompt

Notice the "Low Roof" label is now listed in the *Floor Plans* and *Reflected Ceiling* section of the *Project Browser*.

Add another gable roof:

10. Open the newly created **Low Roof** Floor Plan view.

11. Select the **Roof** tool from the *Design Bar*.

12. Select "**Roof by footprint**" from the pop-up menu.

Before you start the roof you will change the slope (pitch) of the roof.

13. Click the **Properties** button on the *Options Bar*.

14. Change the **rise/12** to **8"**. Click **OK**. (Figure 7-3.5)
    TIP: Make sure "Defines Slope" is selected.

Again, this will make the roof pitch 8/12, which means; for every 12" horizontally the roof will *rise* 8" vertically.

15. Set the overhang to **1'-6"** on the *Options Bar*.

16. You are now prompted to select exterior walls to define the footprint. Select ONLY the three wall segments that define the Living Room area per **Figure 7-3.6**.
    REMEMBER: Pick the exterior side of the walls.

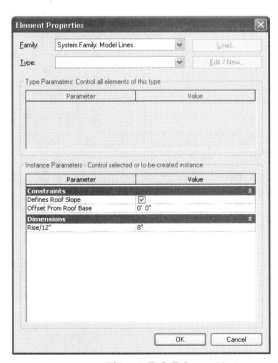

Figure 7-3.5 Properties

You will notice in figure 7-3.6 that there is one section that still needs to be sketched to close the "footprint".

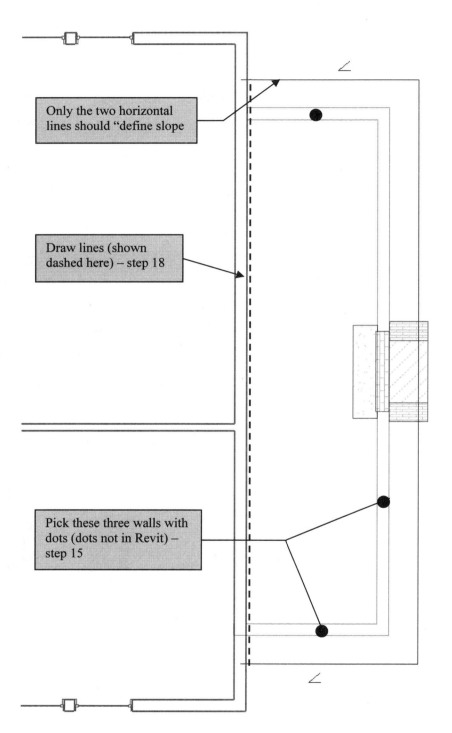

Only the two horizontal lines should "define slope

Draw lines (shown dashed here) – step 18

Pick these three walls with dots (dots not in Revit) – step 15

Figure 7-3.6 Low Roof plan view; adding roof

17. Select the **Lines** tool from the design bar. (Figure 7-3.7)

18. Draw the line to create a complete rectangle (i.e. footprint), making sure you use the snaps to accurately snap to the endpoint of the lines already present. (Figure 7-3.6).

19. Now click **Finish roof** from the *Design Bar*. (Figure 7-3.7)

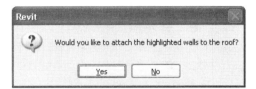

Figure 7-3.8 Prompt

20. Click **Yes** when prompted to attach the highlighted walls to the roof. (Figure 7-3.8)

Figure 7-3.7
Roof sub-tools

You will now see a portion of the roof in your plan view. The cutting plane is 4'-0" above the floor level, so you are seeing the roof thickness in section at 4'-0" above the Low Roof level. See Figure 7-3.9 (next page).

- Switch to an elevation view to see the roof, south elevation, shown in Figure 7-3.10.

- You can also switch to the default 3D view to see the roof in isometric view.

After looking at the roof you have created, switch back to the plan view: **Low Roof**. You will now add a roof over the garage.

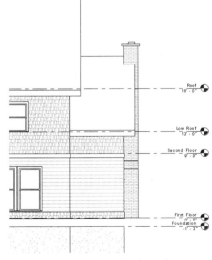

Figure 7-3.10 South elevation

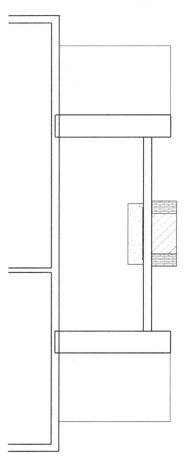

Figure 7-3.9 Low Roof plan view

21. **Zoom in** on the garage area.

22. Select the **Roof** tool; and click *"Roof by Footprint."*

23. With **Defines Slope** checked in the *Options Bar*, pick the two horizontal walls (top and bottom). (Figure 7-3.11)

24. Uncheck **Defines Slope**, and then select the vertical wall on the left and then use the **Line** tool and sketch a line as shown in Figure 7-3.11 to close the footprint. Be sure to use snaps to accurately draw the enclosed area.

25. Select **Finish Roof** from the *Design Bar*. (Figure 7-5)

26. **Click Yes** when prompted to attach the highlighted walls to the roof. (Figure 7-3.8)

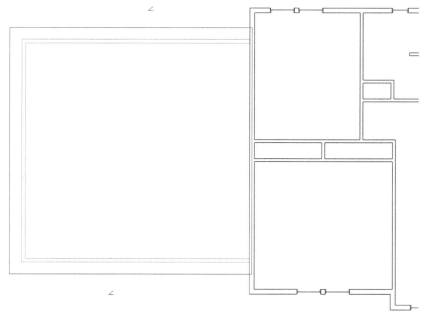

Figure 7-3.11 Low Roof plan view at garage

27. Switch to the **south** elevation view. (Figure 7-3.12)

Figure 7-3.12 South Elevation view: garage roof added

The 3D Model is shown shaded. (Figure 7-3.13)

Revit automatically makes windows and curtain walls transparent.

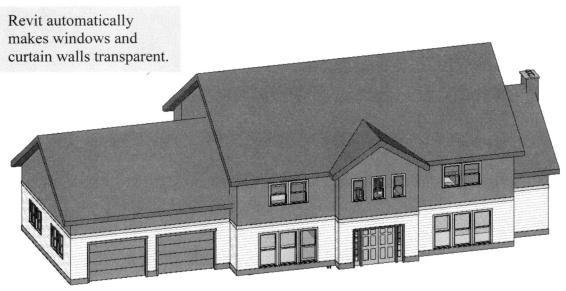

Figure 7-3.13 Shaded model

Changing to a hip roof?

Now that you have drawn the roof, you will see what is involved to modify the gable roof to a hip roof. In the first exercise, in the Lesson, you learned how to create a hip roof from scratch. Here you will modify a previously drawn gable roof into a hip roof.

28. Switch to the **Roof Plan** view.

29. Click on the main roof element to select it.
    TIP: *Select one of the lines at the perimeter of the roof.*

30. Select **Edit** from the *Options Bar*.

31. Selected the vertical line on the left.

32. With the left vertical line selected, check "*Defines Slope*" on the Options Bar.

33. Repeat these steps for the right vertical line.

Your screen should look like Figure 7-3.14.

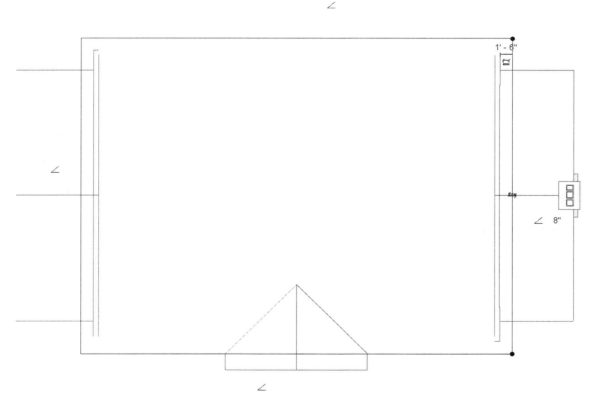

Figure 7-3.14 Roof Plan; vertical lines set to "define slope"

34. Select **Finish Roof** from the Design Bar.

35. Switch to the **South** exterior elevation view.

You now have a hip roof. However, there is a little problem with the lower roofs, they do not properly intersect. This process is exactly the same as the steps taken to join to main entry roof to the second floor roof. You will try this again. Remember, practice makes perfect (or at least better)!

Figure 7-3.15 South Elevation; main roof changed to be a hip roof

36. Switch to the **3D** view.

37. From the *Tools* menu, select **Join/Unjoin Roof** tool.

38. **Join** the lower roof over the garage to the upper roof.
    REMINDER: *Select the edge of the low roof and then select the face of the upper roof (i.e. the face to extend to).*

39. Now **Join** the low roof over the Living Room with the upper roof.
    TIP: *You can also use the Join Roof icon to initiate the command.*

Your model should look similar to Figure 7-3.16.

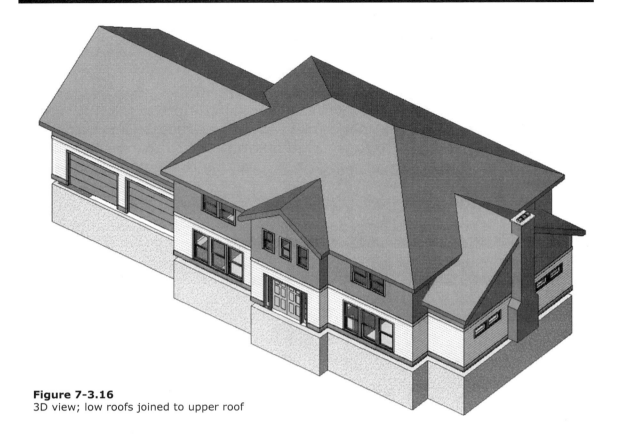

Figure 7-3.16
3D view; low roofs joined to upper roof

Here you can see that the bottom of the roofs, to be joined, do not have to align with each other.

After looking at the two roof options, the client decides he likes the gable option so you will change it back. Do not simply click Undo, the next steps will show you how to Unjoin the roof.

   40. Switch to the **Roof** plan view.

   41. From the *Tools* menu, select **Join/Unjoin**.

This tool, as its name implies, allows you to both *Join* and *Unjoin* roof elements. To *Unjoin*, you simply click on a line that has been created by a *Join* and the entire *Join* will go away.

   42. Select the valley line as shown in **Figure 7-3.17**.

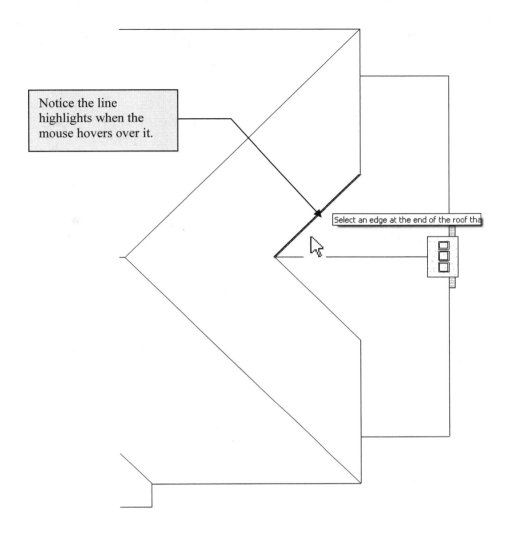

Notice the line highlights when the mouse hovers over it.

Select an edge at the end of the roof tha

Figure 7-3.17
Roof plan; Unjoining the lower roof from the upper roof

As soon as you click the valley line, the roof in unjoined.

43. Repeat this process to *Unjoing* the west lower roof.

44. Change the hip roof back to a gable using the techniques previously covered.

When finished, your 3D view should look like 7-3.13 again.

45. **Save** your project.

Exercise 7-4:
Skylights

This short exercise covers inserting skylights in your roof. The process is much like inserting windows. In fact, Revit lists the skylight types with the window types, so you use the *Window* tool to insert skylights into your project.

Inserting skylights:

You will place the skylights in an elevation view.

1. Load project file **ex7-3.rvt**.

2. Switch to the **North** elevation view.

3. Select the **Window** tool and use the *Load* button to import the *skylight* Family (skylight.rfa) into the current project.

4. Select **Skylight: 24" x 27"** from the type selector.
   (*A 24" wide skylight fits nicely if the trusses are space 24" O.C.*)

You are now ready to place skylights in the roof. Revit will only look for roof objects when placing skylights, so you don't have to worry about a skylight ending up in a wall.

5. Roughly place two skylights as shown in Figure 7-4.1.

Figure 7-4.1 South elevation, skylights added

6. Press **Esc** or click the **Modify** tool to cancel the *Window* tool.

Next, you will want to align the skylights with each other.

7. Switch to the **East** elevation view.

8. Select one of the visible skylights.

9. Select the **Activate Dimensions** button on the Options Bar.

You should now have the skylight selected and see the reference dimensions that allow you to adjust the exact location of the object. Occasionally, the dimension does not go to the point on the drawing that you are interested in referencing from. Revit allows you to adjust where those temporary dimensions point to.

10. Click and drag the grip shown in Figure 7-4.2 to (wait until it snaps) the ridge of the main roof. (Figure 7-4.3).

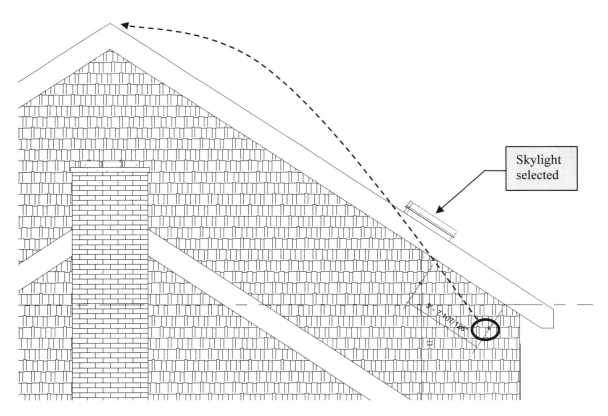

Skylight selected

Figure 7-4.2 East elevation: temp. dimension shown after selecting Activate Dimensions from the Options Bar.

11. Click on the dimension text and change the text to **18'-0"**.
    Note: *This would adjust the position of the skylight relative to the ridge.*

12. Select the other skylight on the **West** elevation and adjust it to match the one you just revised.

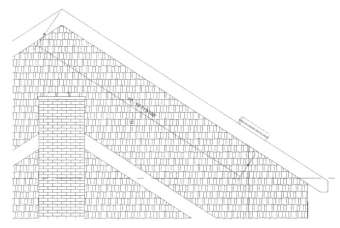

Figure 7-4.3 East elevation: temp dimension witness line adjusted.

Your skylights now align vertically on the roof. The same step would allow you to align the skylight horizontally.

13. Use the temporary dimensions to center the skylight over the second floor windows that are below it, on the north elevation.

14. **Zoom in** on one of the skylights and **click** on the skylight tag (north view).

The number may vary; it is not a problem if it does vary.

15. You should see a symbol appear near the bottom of the tag, Drag on this symbol to move the tag down. (Figure 7-4.4)

16. Position the skylight tag so the tag does not overlap the skylight. (Figure 7-4.5)

This can be done with any tag (i.e. door Number tag, Room tag, etc.).

Figure 7-4.4 Enlarged view of skylight

17. Adjust the other skylights tags, as you reposition these tags you may see a reference line appear indicating the symbol will automatically align with an adjacent symbol.

Remember, the size of the tag is based on the View Scale setting.

Take a minute to look at your shaded 3D view and try changing the view so you can see through the skylight glass into the spaces below. (Figure 7-4.6)

18. **Save** as **ex7-4.rvt**.

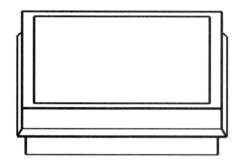

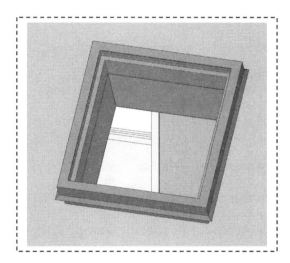

Figure 7-4.6 Shaded skylight view

Figure 7-4.5 Enlarged skylight view — revised

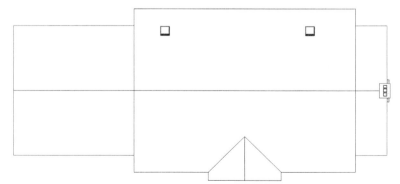

Figure 7-4.7 Roof plan

Self-Exam:

The following questions can be used as a way to check your knowledge of this lesson. The answers can be found at the bottom of this page.

1. You don't have to click *Finish Sketch* when you are done defining a roof. (T/F)

2. The wall below the roof automatically conforms to the underside of the roof when you join the walls to the roof. (T/F)

3. The roof overhang setting is available from the Options Bar. (T/F)

4. To create a gable roof on a building with 4 walls, two of the walls should not have the _____ option checked.

5. Is it possible to change the reference point for a temporary dimension that is displayed while an object is selected? (Y/N)

Review Questions:

The following questions may be assigned by your instructor as a way to assess your knowledge of this section. Your instructor has the answers to the review questions.

1. When creating a roof using the "*create roof by footprint*" option, you need to create a closed perimeter. (T/F)

2. Can the "*defines slope*" setting be changed after the roof is "finished?" (T/F)

3. Skylights need to be rotated to align with the plane (pitch) of the roof. (T/F)

4. Skylights automatically make the glass transparent in shaded views. (T/F)

5. While using the **Roof** tool, you can use the _____ tool from the *Design Bar* to fill in the missing segments to close the perimeter.

6. You use the _____ variable to adjust the vertical position of the roof relative to the current working plane (view).

7. While using the roof tool, you need to select the _____ tool from the *Options Bar* before you can select a roofline for modification.

8. You need to use the _____ _____ to flip the roofline when you pick the wrong side of the wall and the overhang is shown on the inside.

9. The _____ from the Tools menu allows you to extend one roof element over, and into, another.

10. Changing the name of a level tag (in elevation) causes Revit to rename all the corresponding views (plan, ceiling, etc.) if you answer yes to the prompt.

Self-Exam Answers:
1 - F, **2** - T, **3** - T, **4** - defines slope, **5** - Y

Lesson 8
Residence: FLOOR SYSTEMS AND REFLECTED CEILING PLANS::

In this lesson you will learn to create floor structures and reflected ceiling plans.

Even though you currently have floor levels defined, you do not have an object that represents the mass of the floor systems. You will add floor systems with a hole for stair.

Ceiling systems allow you to specify the ceiling material by room and the height above the floor. Once the ceiling has been added it will show up in section views (sections are created later in this book).

Exercise 8-1:
Floor Systems

Similar to other Revit objects, you can select from a few pre-defined object types. You can also create new Families/Types. In your residence you will use a pre-defined floor system for the first and second floors and then create a new type for the basement and garage.

Basement floor, Slab on Grade:
Sketching floors is a lot like sketching roofs (Lesson 7), you can select walls to define the perimeter and draw lines to fill in the blanks and add holes (cut-outs) in the floor object.

1. Open ex7-4.rvt and **Save As ex8-1.rvt**.

2. Switch to the **Basement** floor plan view.

3. From the *Basic* tab on the *Design Bar*, select **Floor**.

4. Click the **Floor Properties** command on the *Design Bar*.

5. Select **Concrete Slab – 4"** from the *Type* drop-down.

6. Click **Edit / New**.

7. Click **Duplicate**.

8. Type **4" Slab on Grade over Vapor Barrier**, then **OK**.

9. Click the **Edit** button next to the *Structure* Parameter.

Here you could change the thickness of the slab, but you will leave it at 4" for the basement.

Next you will add a vapor barrier and sand bed below the slab.

10. Insert a new Layer:
    a. *Function*: **Membrane Layer**
    b. *Material*: **Vapor / Moisture Barriers – Vapor Retarder**
    c. *Thickness*: **0"**

NOTE: *Vapor barriers do have a thickness of course (4mil, 6mil, etc). However, its thickness is typically ignored on drawings.*

11. Add another *Layer*:
    a. *Function*: **Substrate [2]**
    b. *Material*: **Site - Sand**
    c. *Thickness: **4"**
       (Figure 8-1.1)

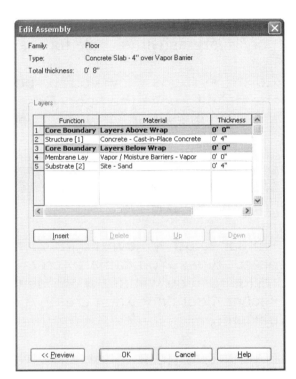

Figure 8-1.1 New Floor System

NOTE: *The sand is sometimes placed below the vapor barrier. The send helps to protect the vapor barrier until the rebar and concrete slab are in place. Additionally, the sand allows the concrete to cure more evenly with less chances of cupping. On the other hand, a vapor barrier directly below the slab helps prevent some floor finishes from delaminating due to moisture in the sand.*

12. Click **OK** to close the open dialog boxes.

13. Select the walls in the **Basement plan** as shown in Fig. 8-1.2.
    TIP: *Select the interior side of the wall; you can use the control arrows if needed (opposite of what you did for the roof).*

14. Click **Finish Sketch**.
    TIP: *If you get any errors, see the next page.*

Common errors finishing a floor sketch:

If you get an error message when trying to finish a floor sketch, it is probably due to a problem with the sketched perimeter lines. Here are two common problems:

Perimeter not closed:

You cannot finish a floor if there is a gap, large or small, in the perimeter sketch. *Fix*: sketch a line to close the loop.

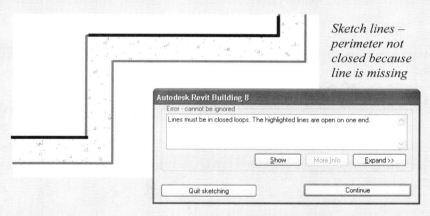

Sketch lines – perimeter not closed because line is missing

Error message after clicking finish sketch

Perimeter lines intersect:

You cannot finish a floor if any of the sketch lines intersect. *Fix*: use Trim to make it so all line endpoints touch.

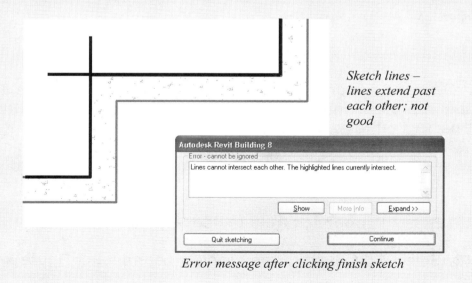

Sketch lines – lines extend past each other; not good

Error message after clicking finish sketch

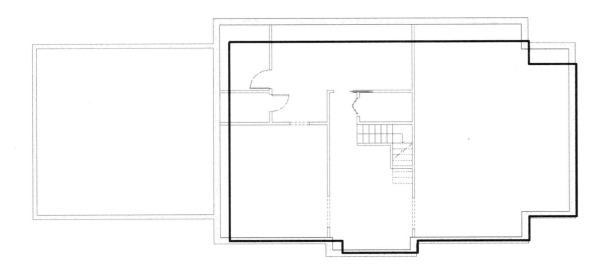

Figure 8-1.2 Basement floor plan; floor slab sketched

15. Click **No** when prompted to join the floor to the walls. (Figure 8-1.3)

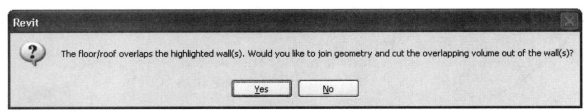

Figure 8-1.3 Basement view Properties

You now have a floor at the basement level. You should see a stipple pattern representing the floor area. You would most likely want to turn that pattern off for a floor plan. You will do that next.

16. In the *Project Browser*, right-click on the **Floor Plan: Basement** view.

17. Select **Properties** from the pop-up menu.

18. Select **Edit**, next to the *Visibility* parameter. (Figure 8-1.4)

19. In the *View Visibility/Graphics window*, click the "plus" sign next to **Floors** in the Visibility column. (Figure 8-1.5)

20. Uncheck **surface pattern**.

21. Click **OK** to close the dialogs.

The stipple pattern is no longer visible.

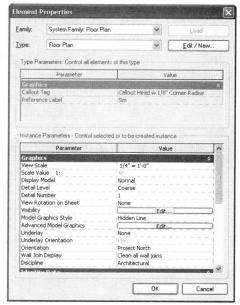

Figure 8-1.4 Basement view Properties

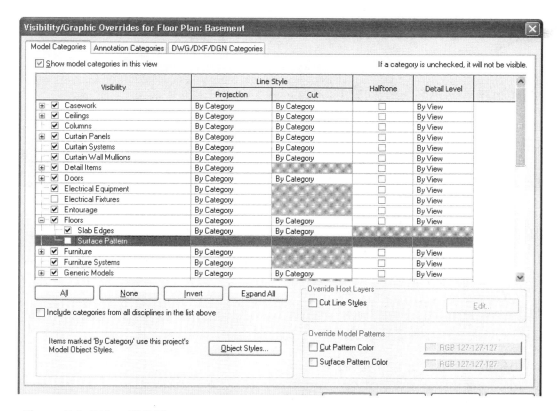

Figure 8.1-5 View Visibility

Garage floor, Slab on Grade:

Now you will sketch the floor slab for the garage area. Because this floor will have vehicles parking on it, you will create a new slab *Type* that is 6" thick.

22. Switch to the **First floor plan** view.

23. Select the **Floor** tool.

24. Using the floor type you just created as a starting point, create a new floor type named **6" Slab on Grade over Vapor Barrier**.
    TIP: *Floor properties, Edit/New, Duplicate*

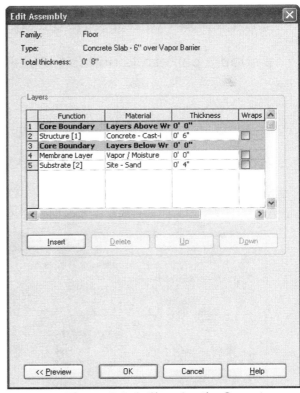

25. Click the **Edit** button next to the *Structure* Parameter.

26. Change the Concrete thickness from 4" to 6". (Figure 8-1.6)

27. Click **OK** to close the open dialog boxes.

Figure 8.1-6 Changing the Concrete Slab thickness

28. With *Pick Walls* selected on the *Design Bar*, select the interior side of the four walls surrounding the garage. (Figure 8.1-7)

29. Click **Finish Sketch**.

You now have a floor slab in the garage area. You will notice that the floor element has a stipple pattern, even though you turned that stipple pattern off in the Basement view. Remember, the Visibility settings are controlled by view, not project wide. You will turn this off later.

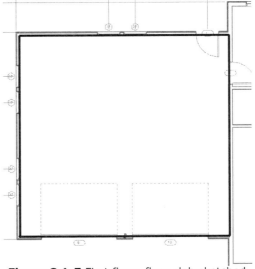

Figure 8.1-7 First floor: floor slab sketched

First floor (wood joists):

30. Use the wall type *Wood Joist 10" - Wood Finish* as a starting point to create a new wall type named **Wood Joist 14" - Wood Finish**.

31. Change the *Structure* settings per the following image. (Figure 8-1.8) *FYI*: *This has all been covered previously.*

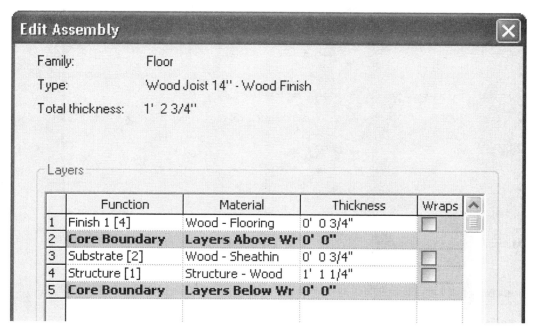

Figure 8.1-8 Floor properties

Creating the first and second floors will be a little more involved than was the basement and garage floors. This is because the upper floors require openings. For example, you need to define the openings for the stair. Revit makes the process very simple.

You should still be in the *Floor* tool.

32. On the *Options Bar*, check "**extend into wall (to core)**."

> **FYI:** The "*extend into wall (to core)*" option will extend the slab under your wall core (wood stud is the core in our example), and go under the gypsum wall board.

33. Select the exterior walls indicated in Figure 8-1.9, select the exterior side of the wall; use the "flip control" arrows if needed.

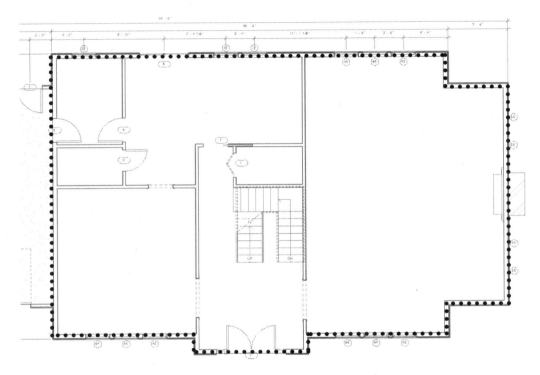

Figure 8-1.9 First Floor – floor system sketched

Next you will define the hole in the floor for the stair that leads down to the basement. You will need to use the *Lines* sub-tool to define this area.

34. While the Floor tool is still active, click on the **Lines** tool from the *Design Bar.*

35. **Zoom In** to the stair area.

36. Select the **rectangle** icon on the *Options Bar.*

37. Draw a <u>rectangle</u> defining the stair opening; use Revit's snaps to accurately pick the points as shown in **Figure 8-1.10**.

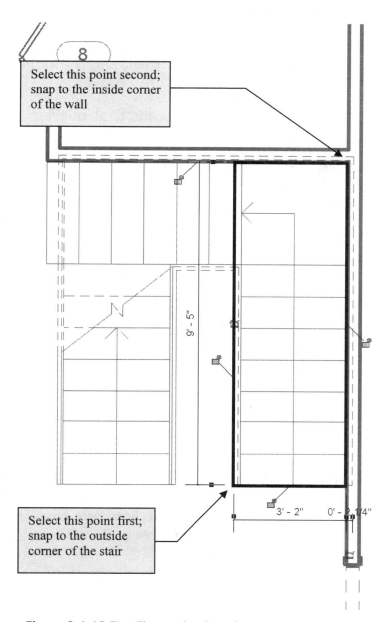

Figure 8-1.10 First Floor – sketching floor opening

FYI: *This floor opening has a big problem. You will see this problem when you get to the building sections exercise later in this book.*

38. Click **Finish Sketch.**

Next you will get two prompts; one asks if you want the walls below (whose tops are set to terminate at the first floor) to attach to the new floor: the other asks if you want the floor to attach to the exterior walls. You will answer yes to both.

39. Click **Yes** to the prompt "*Would you like the walls that go up to this floor's level to attach to its bottom?*" (Figure 8-1.11)
    FYI*: This will allow the walls to dynamically change if the floor elevation or thickness is changed.*

40. Click **Yes** for the prompt to join the walls that overlap the floor system. (Figure 8-1.12)

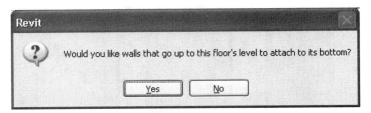

Figure 8-1.11 Join walls below to new floor prompt

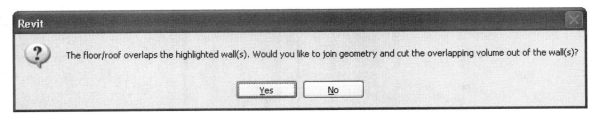

Figure 8-1.12 Join walls that overlap the new floor prompt

Now that the floor is placed, you should see a dense horizontal line pattern that represents the finished wood floor. This would be nice for presentation drawings, but, like the basement, you will turn the floor hatch off so the floor plan is not too cluttered.

41. Change the first floor's *Visibility* to turn off the floor hatch.

Copy/Paste the first floor to the second:

That completes the first floor system. Next you will copy the floor you just created to the second floor. It would probably be easier to just create a floor from scratch, but you will get more practice editing existing (i.e. already drawn) elements.

42. Select the floor you just created (in the first floor plan view) and select **Copy to Clipboard** from the *Edit* pull-down menu.

TIP: <u>Selecting objects that overlap</u>, like the exterior walls and the edge of floor system, may require the use of the **TAB key**. The only way to select a floor object is by picking its edge. Revit temporarily highlights objects when you move your cursor over them. But, because the floor edge may not have an "exposed" edge to select, you will have to toggle through your selection options for your current cursor location. With the cursor positioned over the edge of the floor (probably with an exterior wall highlighted), press the TAB key to toggle through the available options. A tool-tip will display the objects; when you see *floors*: *floor:floor-name*, click the mouse.

---

43. Switch to the **Second Floor**.

44. Pick **Edit → Paste Aligned → Current View**.

You now have a copy of the first floor's "floor system" in the second floor plan view; keep in mind that all the views are looking at a single 3D model. You will edit the sketch lines of the second floor slab to make the stair hole larger and remove a portion of the floor that extends past the east exterior wall.

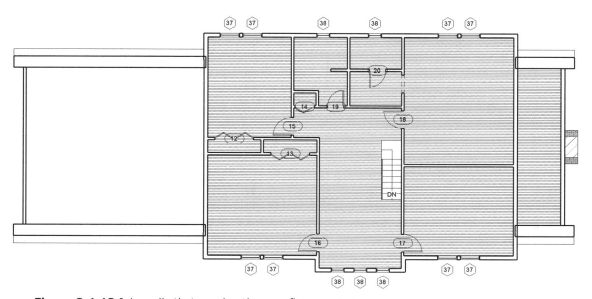

Figure 8-1.13 Join walls that overlap the new floor prompt

45. Select the Floor element in the second floor plan view.

46. Click **Edit** from the *Options Bar*.

You are now in sketch mode, where you can change the perimeter / holes for the selected floor element. You can select lines and delete them, sketch new lines, and use Trim to modify lines.

47. Modify the lines at the far right, using *Delete* and *Trim* so your sketch likes like Figure 8-1.14.

48. *Move* the west vertical line **4'-10"** to the west (the horizontal lines should automatically extend), as shown in Figure 8-1.14
   a. The overall width of the stair opening should be 8'-0".

49. Click **Finish Sketch** to recreate the floor element.

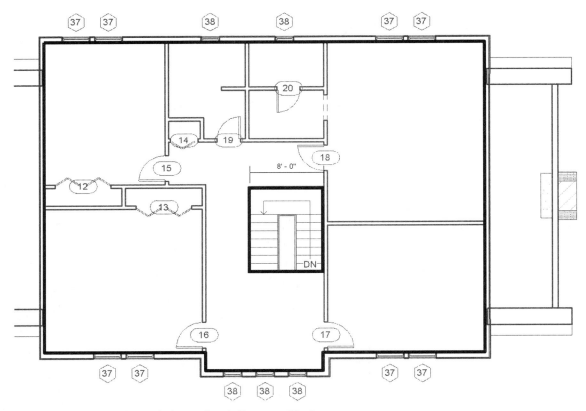

Figure 8-1.14 Second Floor: sketch lines modified

Now that Revit is recreating the floor, it now notices the walls below and at the exterior; so you get the "join" prompts.

50. Click **OK** to the two "join" prompts.

Your second floor plan view should look like Figure 8-1.15.

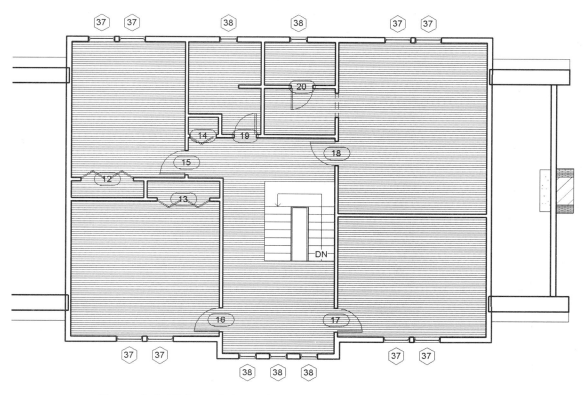

Figure 8-1.15 Second Floor: sketch lines modified

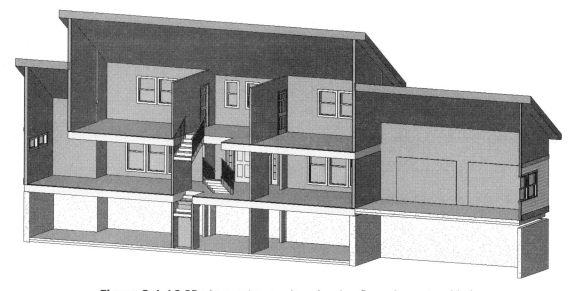

Figure 8-1.16 3D view: cutaway view showing floor elements added

51. Change the second floor's *Visibility* to turn off the floor hatch.

Turning off surface pattern at the source:
If you do not want a portion of your plan to show a surface pattern, you might want to remove the surface pattern associated with the floor system. That way you could leave the surface pattern setting turned on. Then you can add *Regions* to portions of plan views.

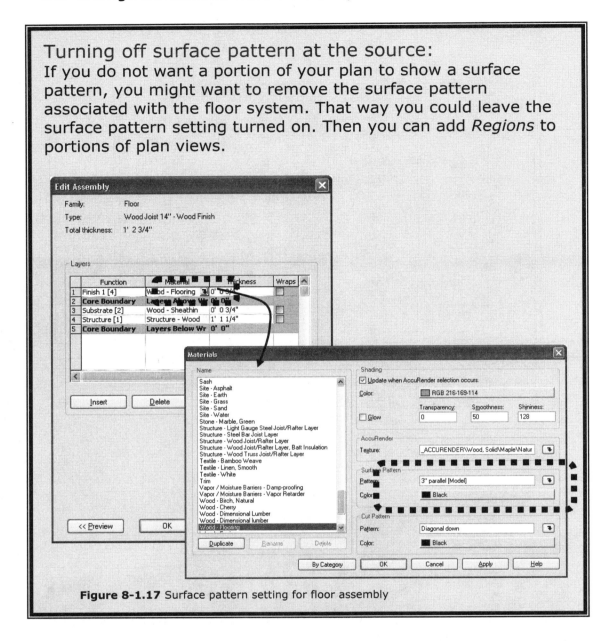

Figure 8-1.17 Surface pattern setting for floor assembly

52. Save your project as **ex8-1.rvt**.

Exercise 8-2:
Ceiling Systems (Susp. ACT & Gypsum Board)

This lesson will explore Revit's tools for drawing reflected ceiling plans. This will include drawing different types of ceiling systems.

Traditionally, residential projects do not typically include reflected ceiling plans. The ceilings are generally straight forward, with all required information coming from a Room Finish Schedule and Sections. However, in Revit, you will sketch the ceiling profile in the ceiling plan views (similar to how you drew the floor systems). You want to add the ceilings so they show up in sections. So, seeing as you have ceiling plans, why not include them in the construction documents set? Your design intent will be clear; and you can make sure the lighting design is coordinated and laid out the way you want it.

The use of suspended acoustical ceiling tile in residential projects is somewhat limited. However, it is occasionally used so it will be covered here. In this lesson you will add it to the second floor office and a family room in the basement.

Suspended Acoustical Ceiling Tile System:

1. Open ex8-1.rvt and Save As **ex8-2.rvt**.

2. Switch to the **Second Floor** ceiling plan view, from the *Project Browser.*

Notice the doors and windows are automatically "turned off" in the ceiling plan views. Actually, the ceiling plan views have a cutting plane similar to floor plans; except they look up rather than down. You can see this setting by right-clicking on a view name in the *Project Browser* and selecting *Properties*, and then selecting **View Range**. The default value is 7'-6". You might increase this if, for example, you had 10'-0' ceilings and 8'-0" high doors. Otherwise, the doors would show because the 7'-6" cutting plane is below the door height. (Figure 8-2.1)

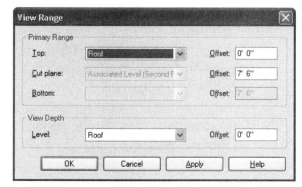

Figure 8-2.1 Properties: View Range settings

3. From the *Modeling* tab on the *Design Bar*, select **Ceiling**.

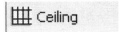

You have 4 ceiling types (by default) to select from. (Figure 8-2.2)

Figure 8-2.2 Ceiling: Options Bar

4. Select **Compound Ceiling: 2'x4' ACT Grid**.

Next you will change the ceiling height. The default setting is 8'-0" above the current level. You will change the ceiling height to 8'-4", this will give about 6 ½" of space above the ceiling to the bottom of the rafters (for recessed light fixtures and any required wiring). This setting can be changed on a room by room basis.

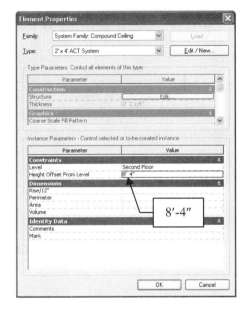

5. Click the **Properties** button. (Figure 8-2.2)

6. Set the *Height Offset From Level* setting to **8'-4"**. (Figure 8-2.3)

Figure 8-2.3 Ceiling: Properties

You are now ready to place ceiling grids. This process can not get much easier, especially compared to other CAD programs.

7. Move your cursor anywhere within the office in the lower-right corner of the building. You should see the perimeter of the room highlighted.

8. Pick within the large room; Revit places a grid in the room. (Figure 8-2.4)

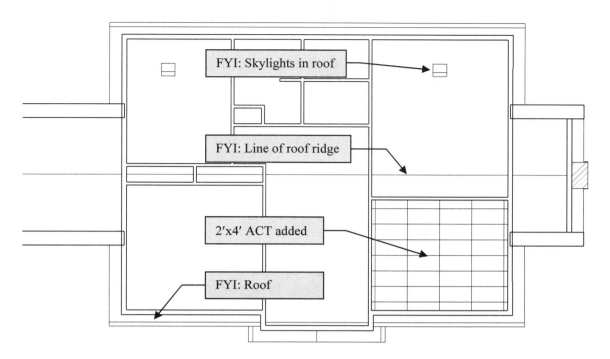

Figure 8-2.4: Second Floor Ceiling Plan view: 2'x4' suspended acoustical ceiling tile added to office

You now have a 2x4 ceiling grid at 8'-4" above the floor (second floor in this case). Later in the book, when you get to the exercise on cutting sections, you will see the ceiling with the proper height and thickness.

Next you will add the same ceiling system to the family room in the basement. But, you will notice the ceiling plan category in the Project Browser does not have a Basement view.

Most basements do not have any finished ceilings, so that view was not included in the template from which you started. Next you will create a view for this.

> 9. From the *View* pull-down, select **New → Ceiling Plan**.
>
> 10. Select the **Basement** view from this list and then click **OK**.
>
> 11. Switch to the **Basement** ceiling plan view.

Your view should look a little funny, no walls. You will have to adjust the View Range settings.

12. Set the *View Range* to **6'-6"**.

You should now see the walls and doors. The doors show because the cut plain is lower than the top of the door. You will ignore that for now.

13. Use the **Ceiling** tool to place a 2'x 4' acoustic ceiling tile in the Family Room (the large room to the far east). See the following notes (Figure 8-2.5):
    a. You will probably get a symbol at the cross-hairs that indicate you cannot place a ceiling anywhere (the symbol is a circle with a slash through it). This has to due with the ceiling height.
    b. Via *Properties*, change the ceiling height to **7'-0"**.
    c. You should be able to place the ceiling system.
    d. Revit automatically rotates and centers the ceiling grid within each room.

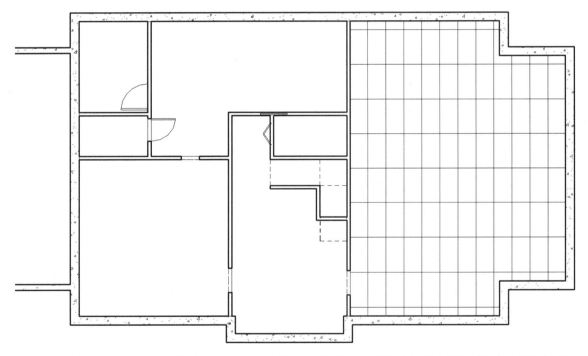

Figure 8-2.5: Basement Ceiling Plan view: 2'x4' suspended acoustical ceiling tile added to Family Rm.

When you place a ceiling grid, Revit centers the grid in the room. The general rule-of-thumb is you should try to avoid reducing the tile size by more than half of a tile. You can see in Figure 8-2.5 that the east and west sides look ok. However, the north and south sides are small slivers. You will adjust this next.

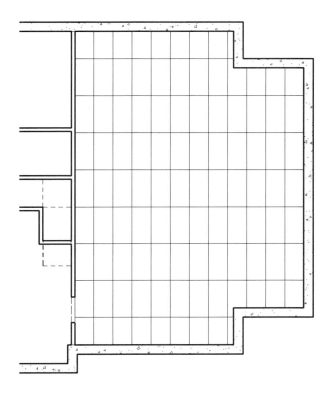

Figure 8-2.6 Basement: Ceiling Grid moved

14. Select **Modify** from the *Design Bar.*

15. **Select** the ceiling grid (only one line will be highlighted).

16. Use the **Move** tool to move the grid 24" to the north. (Figure 8-2.6)

Modifying the Suspended Acoustical Ceiling Tile System:

Making modifications to the grid is relatively easy. Next, you will adjust the ceiling height and rotate the grid.

17. In the Basement view, select the ceiling grid in the Family Room.

18. With the Family Room grid selected, click the **Properties** button from the *Options Bar.*

19. Change the height to **7'-2"**, then click **OK**. *(Every inch counts!)*

20. Again, with the grid still selected, use the **Rotate** tool to rotate the grid **30 degrees**.

TIP: When using the **Rotate** tool ⬜ Rotate you need to pick two points. The first point is your reference line. The second point is the number of degrees off that reference line. In this example, try picking your first point to the right as a horizontal line. Then move the cursor counter-clockwise until 45 degrees is displayed. After you click your second point you can click on the angle dimension and type a more precise value, if necessary.

Your drawing should look similar to **Figure 8-2.7**. Notice how the perimeter of the room acts like a cropping window when you use move and/or rotate on a ceiling grip.

Deleting a ceiling grid:

When selecting a ceiling grid, Revit only selects one line. This does not allow you to delete the ceiling grid. To delete: hover cursor over a ceiling grid line and press the TAB key until you see the ceiling perimeter highlight, then click the mouse. The entire ceiling will be selected. Press Delete.

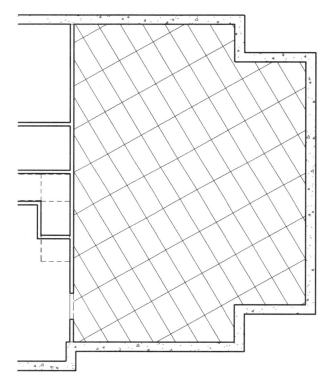

Figure 8-2.7 Basement: Modified Ceiling

Next, you will look at drawing gypsum board (or drywall) ceiling systems. The process is identical to placing the grid system. Additionally, you will create a ceiling type.

Gypsum Board Ceiling System:

You will create a new ceiling type for a gypsum board ceiling. To better identify the areas that have a gyp. bd. ceiling, you will set the ceiling type to have a stipple pattern. This will provide a nice graphical representation for the gyp. bd. ceiling areas. The ceiling you are about to create would lean more towards a commercial application. However, it will give you a better understanding of how to create custom ceilings. You will add this ceiling to a room in the basement; assume it is being installed below HVAC ductwork, to finish the room.

21. From the *Settings* pull-down menu, select **Materials**.
    This is the list of materials you select from when assigning a material to each layer in a wall system, etc.

22. Select *Finishes - Interior - Gypsum Wall Board* in the Name list and then click the **Duplicate** button and enter the name: **Finishes - Interior - Gypsum Ceiling Board**.

23. In the *Surface Pattern* area, pick the arrow button and select **Gypsum-Plaster** from the list, and then click **OK** twice. (Figure 8-2.8)

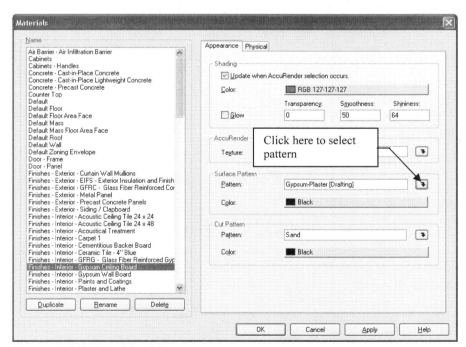

Figure 8-2.8 Materials dialog

The *Surface Pattern* setting is what will add the stipple pattern to the gypsum board ceiling areas. With this set to *none*, the ceiling has no pattern (like the basic ceiling type).

Thus, if you wanted Carpet 1 finish to never have the stipple hatch pattern, you could change the surface pattern to none via the Materials dialog and not have to change each views visibility override.

24. From the *Modeling* tab on the Design Bar, select **Ceiling**.

25. Click the **Properties** button. (Figure 8-2.2)

26. Set the Type to **GWB on Mtl. Stud**. *FYI: You are selecting this because it is similar to the ceiling you will be creating.*

27. Click the **Edit / New** button.

28. Click *Duplicate* and type the name **Susp GB on Metal Stud**.

29. Select **Edit** next to the *Structure* parameter.

30. Set the Values as follows: (Figure 8-2.9)
    a. **1 ½" Mtl. Stud**
    b. **¾" Mtl. Stud**
    c. **Finishes – Interior – Gypsum Ceiling Board**
       (This is the material you created in step 21.)

31. Click **OK** three times.

FYI: The ceiling assembly you just created represents a typical suspended gypsum board ceiling system. The Metal Studs are perpendicular to each other and suspended by wires, similar to an ACT (acoustical ceiling tile).

You are now ready to draw a gypsum board ceiling.

32. Make sure **Susp GB on Metal Stud** is selected in the *Type Selector* on the *Options Bar.*

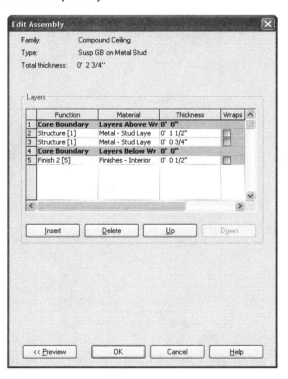

Figure 8-2.9 New ceiling – Edit assembly

33. Set the ceiling height to **7'-0"**.

34. Pick the lower left room as shown in **Figure 8-2.10**.

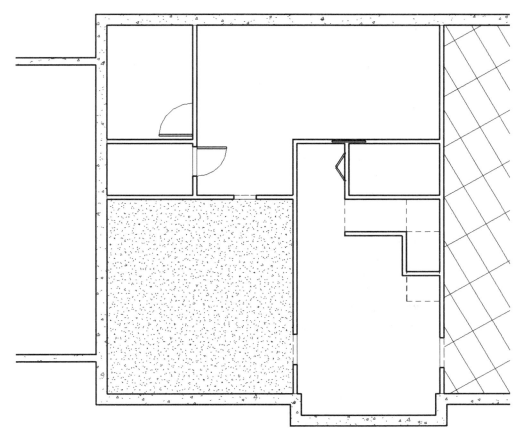

Figure 8-2.10 Gypsum Board Ceiling

You now have a gypsum board ceiling at 7'-0" above the basement floor slab.

Adding a bulkhead:

Next, you will draw a ceiling in the area at the bottom of the steps. However, you cannot simply pick the room to place the ceiling because Revit would fill in the stair area. First, you will need to draw a bulkhead at the bottom of the steps. A bulkhead is a portion of wall that hangs from the floor above and creates a closed perimeter for a ceiling system to die into. The bulkhead will create a perimeter that the Ceiling tool will detect for the proper ceiling placement.

35. While still in the Basement Reflected Ceiling Plan view, select the **Wall** tool.

36. Click **Properties** from the *Options Bar.*

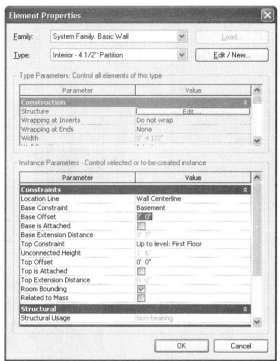

37. Set the *wall type* to: **Interior - 4 1/2" Partition** and the *Base Offset* to **7'-0"**. *(This will put the bottom of the wall to 7'-0" above the current floor level, basement in this case.)* (Figure 8-2.11)

38. Set the Top Constraint to: **Up to level: First Floor**. (Figure 8-2.11)

TIP: The next time you draw a wall you will have to change the Base Offset back to 0'-0" or your wall will be 7'-0" off the floor.

Figure 8-2.11 Bulkhead (wall) properties

39. **Draw the bulkhead**; make sure you snap to the adjacent walls (Figure 8-2.12)

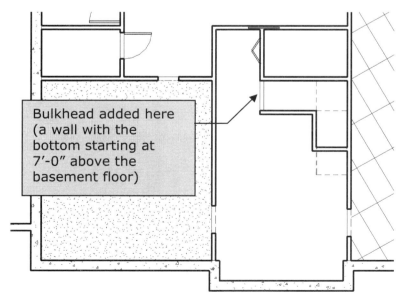

Bulkhead added here (a wall with the bottom starting at 7'-0" above the basement floor)

Figure 8-2.12 Bulkhead drawn

40. Select the **Ceiling** tool.

Next you will sketch the perimeter of the ceiling, just like you did for the floor slabs in the previous lesson.

41. Select **Sketch** from the *Options Bar*.

42. Sketch a line around the perimeter of the room, as shown in Figure 8-2.13 *TIP*: *Remember to use the chain option.*

43. Click **Finish Sketch**.

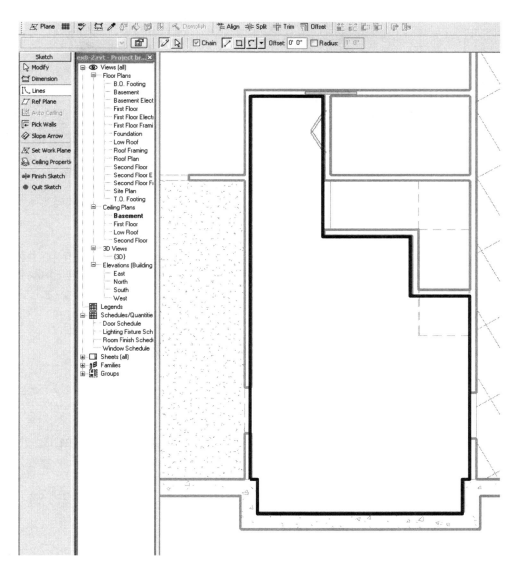

Figure 8-2.13 Sketching Ceiling

Your ceiling should look like Figure 8-2.14. You will see what is happening here a little better in the sections exercise.

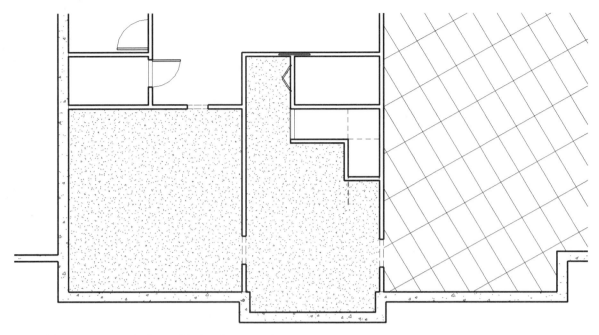

Figure 8-2.14 Ceiling added at bottom of steps

Modifying a floor system to include the ceiling:

Next you will modify the floor systems (already in place) to have gypsum board attached to the bottom of the joists. This is what is typically done in a residence, unless room is needed above the ceiling for ductwork, etc.

44. Switch to the **Second Floor** plan view.

45. Select the floor. *TIP: Clicking near the stair opening may be the easiest way to select the floor.*

46. Create a Duplicate floor from the one selected, name it: **Wood Joist 14" – Wood Finish – Gyp Bd Ceiling**

47. Edit the floor structure via properties. (Figure 8-2.15)

    a. Insert a new Layer at the bottom of the floor assembly
    b. Set the *Function* to: **Finsh 1 [4]**
    c. Set the *Material* to: **Finishes - Interior - Gypsum Ceiling Board**
    d. Set the *Thickness* to: **½"**

Why did you create a new floor type?

You needed to create a new floor type for the second floor because the first floor does not need gypsum board on the underside. When you modify a Family Type, all instances of that type are changed throughout the project. This is very powerful when you actually want to change several instances of a type. However, if you modify a type and only want the selected one to change, you're in for a big surprise when you look the other instances of that type elsewhere in your project (because they will all change). That is why you had to create a new floor type (because that same floor type is all placed on the first floor).

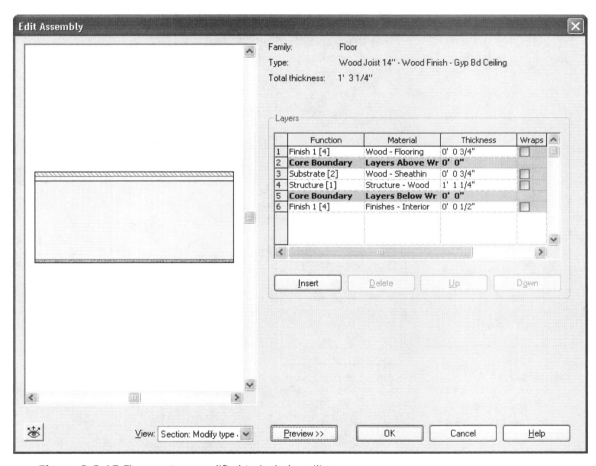

Figure 8-2.15 Floor system modified to include ceiling

48. Close the Open dialog boxes.

Now switch to the First Floor ceiling plan (Figure 8-2.16) and notice that the gypsum ceiling board surface pattern is showing everywhere the floor occurs (if not, make sure the second floor system has been changed to the new type). That takes care of the ceiling for the first floor.

49. Switch to the Second Floor Ceiling plan and add your **Susp GB on Metal Stud** at 8'-0" high to the remaining rooms (see Figure 8-2.17). *FYI: Normally you would attach the gypsum board to the bottom of the truss.*

50. **Save** your project as **ex8-2.rvt**.

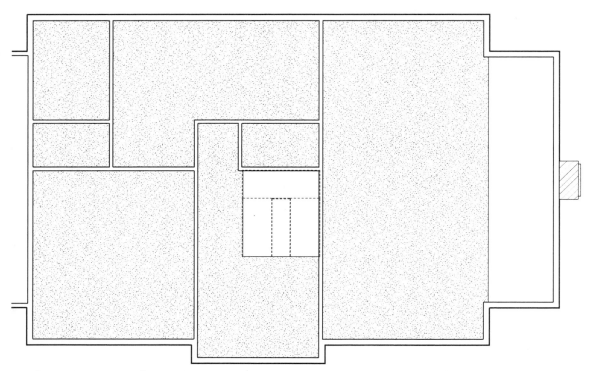

Figure 8-2.16 First floor ceiling plan: gypsum board added to bottom of floor structure

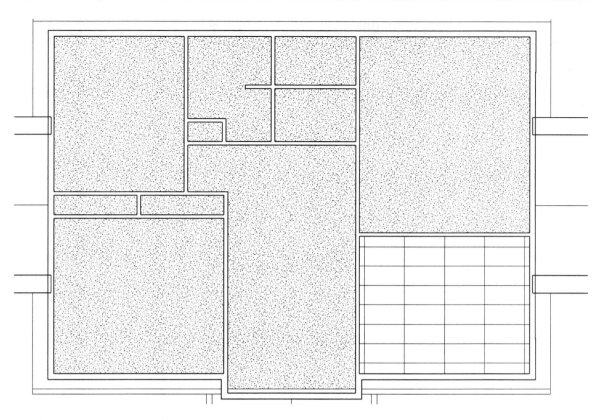

Figure 8-2.17 Second floor ceiling plan: gypsum board ceiling system suspended at 8'-0" AFF

Exercise 8-3:
Placing Light Fixtures

In this exercise, you will learn to load and place light fixtures in your reflected ceiling plans.

Loading Components:

Before placing fixtures, you need to load them into your project.

1. Select **Component** from the *Modeling* tab on the *Design Bar.*

2. Select **Load** on the *Options Bar.* (Figure 8-3.1)

Figure 8-3.1 Component; Options Bar

3. Double-click the *Lighting Fixtures* folder, and then Double-click **Troffer - 2x4 Parabolic.rfa**. (Figure 8-3.2)

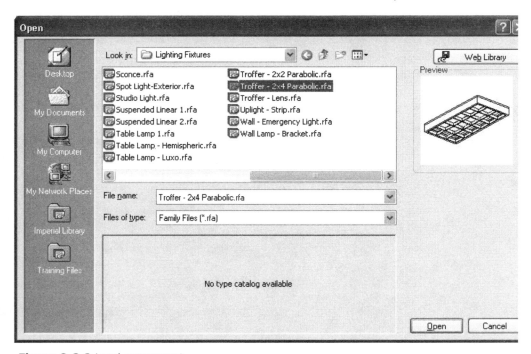

Figure 8-3.2 Load component

Placing instances of components:

You are now ready to place the fixtures in your ceiling plans.

4. With **Component** still selected from the *Design Bar*, pick **Troffer - 2' x 4' Parabolic: 2'x4' (2 Lamp)** from the *Type Selector* drop-down on the *Options Bar.*

5. On the **Second Floor ceiling plan view**, place fixtures as shown in **Figure 8-3.3**.

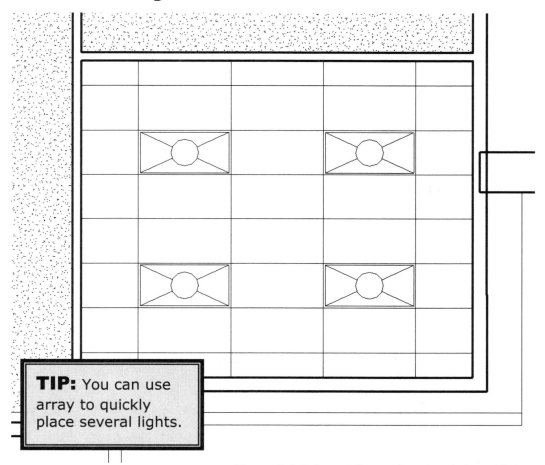

TIP: You can use array to quickly place several lights.

Figure 8-3.3 Second floor ceiling plan; lights added

You may have to use the *Move* command to move the fixture so it fits perfectly in the ACT grid.

6.   Switch to the **Basement** floor plan view.

7.   Now place another **2x4 light fixture** as shown in **Figure 8-3.4**.

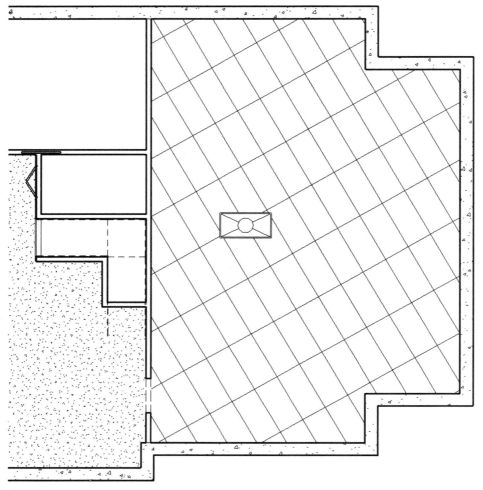

Figure 8-3.4 Basement ceiling plan; light fixture added

Notice the fixture does not automatically orientate itself with the ceiling grid. There may be an occasion when you want this.

Also, notice the light fixture hides a portion of the ceiling grid. This is nice because the grid does not extend through a light fixture.

8. Use **Rotate** and **Move** to rotate the fixture to align with the grid. (Figure 8-3.5) *TIP: You can use the normal rotate and move tools in conjunction with the snaps.*

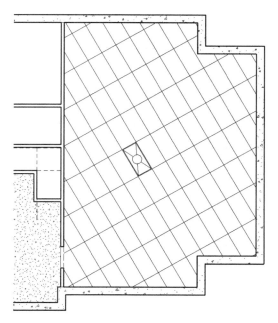

Figure 8-3.5 Basement ceiling plan; light fixture moved/rotated into place

9. Once you have one fixture rotated, it is easier to use the copy tool and the snaps to add rotated light fixtures. **Copy** the light fixture to match the layout in **Figure 8-3.6**.

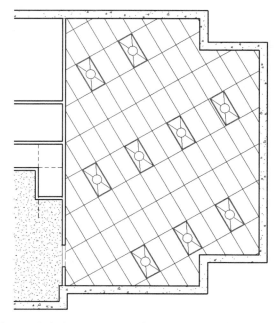

Figure 8-3.6 Basement ceiling plan; light fixture copied

10. **Save** your project as **ex8-3.rvt**.

These 2x4 light fixtures have Accurender settings already setup. That means the room will automatically have lights when you get to creating a photo-realistic rendering using Accurender (which is built into Revit).

The example image below shows a reflected ceiling plan with supply and return diffusers added. These are not used to often in residential work. They are loaded similar to the light fixture via component tool.

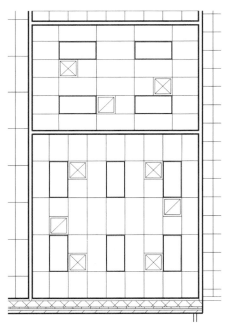

Figure 8-3.7 Example image; supply and return diffusers added

Reflected Ceiling Plan Symbols:

Revit provides many of the industry standard symbols necessary in drawing reflected ceiling plans (RCP). As shown in Figure 8-28, supply air is represented with an X and return air has a diagonal line. It is typical to have a RCP symbol legend showing each symbol and material pattern and list what each one represents. The new Legend feature makes this a snap!

Component Properties

If you want to adjust the properties of a component, such as a light fixture, you can browse to it in the Project Browser and right click on it *(notice the right click menu also has the option to select all instances of the item in the drawing)* and select Properties. You will see the dialog below for the 2x4 (2 Lamps).

You can also click duplicate and add more sizes (e.g. 4'x4' light fixture).

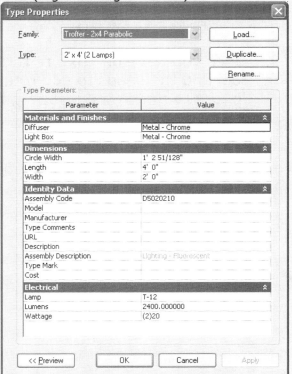

You can also select an inserted component and click the Properties button on the Options Bar for additional properties for that particular instance.

Exercise 8-4:
Annotations

This short section will look at adding notes to your RCP.

Adding Annotations:

1. Select the **Text** tool from the *Design Bar.*

2. Pick **Text: 3/32" Arial** from the *Type Selector*. (Figure 8-4.1)

3. Select the **Leader** (arc) button circled in **Figure 8-4.1**.

Figure 8-4.1 Text; options bar

Next, you will add a note indicating that a room in the basement does not have a ceiing. First you will draw a leader, and then Revit will allow you to type the text.

4. Add the note "**NO CEILING - TYPICAL**" shown in **Figure 8-4.2**.

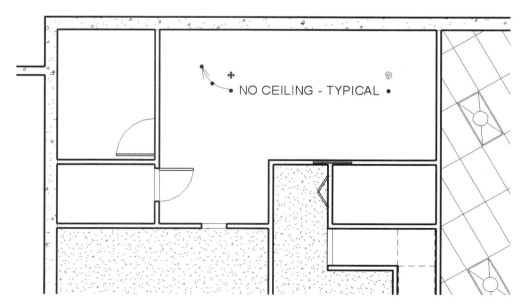

Figure 8-4.2 Text with ARC leader

Notice that immediately after adding the text or while it is selected, you see the *Move* icon, the *Rotate* icon and *Grips* to edit the arrow/leader.

Adding text styles to your project:

You can add additional text styles to your project. Some firms prefer a font that has a hand lettering look and others prefer something like the Arial font. These preferences can be saved in the firms template file so they are consistent and always available. You will add a test style next.

5. Click on the **Text** tool.

6. Select **Properties** on the *Options Bar.*

7. Next, click **Edit/New**.

8. Select **Duplicate** and enter **1/4" outline text**. (Figure 8-4.3)

9. Next, make the following adjustments to the *Type Properties*. (Figure 8-4.4)
    a. Text Font: **Swis721 BdOul BT**
    b. Text Size: **1/4"**

Figure 8-4.3 New text name

NOTE: *You can use any Windows True-Type font. If you do not have this font, select another that best matches (see 8-4.5 below).*

The text size you entered in step 9 is the size of the text when printed. Revit automatically scales the text, in each view, based on the *View Scale* setting. For example; if you are adding 3/32" text to an 1/8" floor plan view, the text placed will be 9" tall. If you add text to a 1/4" plan view, the text will be 4 1/2" tall. When both the 1/4" and 1/8" plan views are placed on a Sheet, the text for each is 3/32" tall. Here is what happens: adding text to the ¼" plan causes the 3/32" text to be scaled up 48 times, and when the plan view is placed on a *Sheet*, the entire view (including the text), is scaled down 1/48 times; thus, the text makes a full loop back to being 3/32" tall.

If you change the [view] scale of the drawing, the text size will automatically change, so the text is always the correct size when printing. It is best to set the drawing to the correct scale first. As changing the *View Scale* can create a lot of work; repositioning resized text that may be overlapping something or too big for a room.

10. Select **OK** to close the open dialog boxes.

You should now have the new text style available in the *Type Selector* on the *Options Bar*.

11. Use the new text style to create the text shown below. (Figure 8-4.5)
    TIP: *Select the text only options on the Options Bar.*

When typing text, you simply click somewhere in the view to complete the text string; pressing *Enter* just adds more lines.

12. Save as **ex8-4.rvt**.

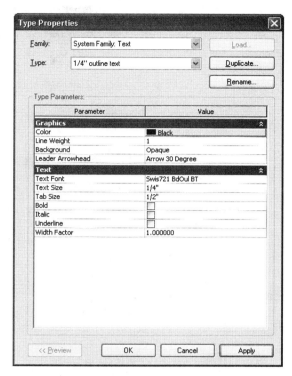

Figure 8-4.4 New text properties

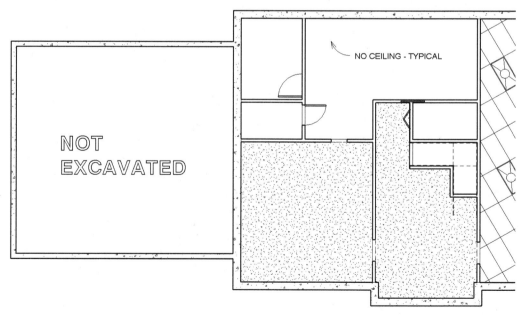

Figure 8-4.5 New text style added

Self-Exam:

The following questions can be used as a way to check your knowledge of this lesson. The answers can be found at the bottom of this page.

1.   You must pick Walls to define floor areas. (T/F)

2.   Use the Ctrl key to cycle through the selection options. (T/F)

3.   When you add a floor object in plan view, the floor does not show up right away in the other views; i.e. 3D, Sections, Etc. (T/F)

4.   You use the _____ tool; if you need to add a new product, like exterior plaster, so you can add it to wall types and other systems.

5.   You have _____ different types of leader options with the text tool.

Review Questions:

The following questions may be assigned by your instructor as a way to assess your knowledge of this section. Your instructor has the answers to the review questions.

1.   It is not possible to create new text styles. (T/F)

2.   You can add additional light figure sizes to the group as required. (T/F)

3.   The light fixtures automatically turn to align with the ceiling grid. (T/F)

4.   You can adjust the ceiling height room by room. (T/F)

5.   To create a hip roof, all four "sketch lines" should be Slope Defining. (T/F)

6.   It is not possible to place ceiling objects in what part of a ceiling plan

     _____. *TIP: While Revit is showing this:* $\oslash$

7.   Use the _____ tool if the ceiling grid needs to be at an angle.

8.   Use the _____ tool to adjust the ceiling grid location if a ceiling tile is less than half its normal size.

9.   Use the _____ tool to adjust whether an object's surface pattern is displayed (i.e. the stipple for the gypsum board ceiling).

10. What is the current size of your project (after completing exercise 8-4)?

     _____ MB.

Self-Exam Answers:
1 - F, **2** - F, **3** - F, **4** - Material, **5** - 4

Lesson 9
Residence: ELEVATIONS::

This lesson will cover interior and exterior elevations. The default template you started with already has the four main exterior elevations set up for you. You will investigate how Revit generates elevations and the role the elevation tag plays in that process.

Exercise 9-1:
Creating & Viewing Parametric Exterior Elevations

Here you will look at setting up an exterior elevation and how to control some of the various options.

Setting up an exterior elevation:

Even though you already have the main exterior elevations set, you will go through the steps necessary to set one up. Many projects have more than four exterior elevations, so all exterior surfaces are elevated.

1. Open your project, ex8-4.rvt, and **Save As ex9-1.rvt**.

2. Switch to your **First Floor** plan view.

3. Select **Elevation** ⊹ Elevation , from the *View* tab on the *Design Bar.*

    TIP: If you do not see the view tab on the Design Bar, right-click on one of the visible tabs and select View from the pop-up menu (displaying all available design bar tabs).

4. Place the temporary elevation tag in plan view as shown in Figure 9-1.1.

    NOTICE: As you move the cursor around the screen, the elevation tag automatically turns to point at the building.

You now have an elevation added to the *Project Browser* in the *Elevations* category.

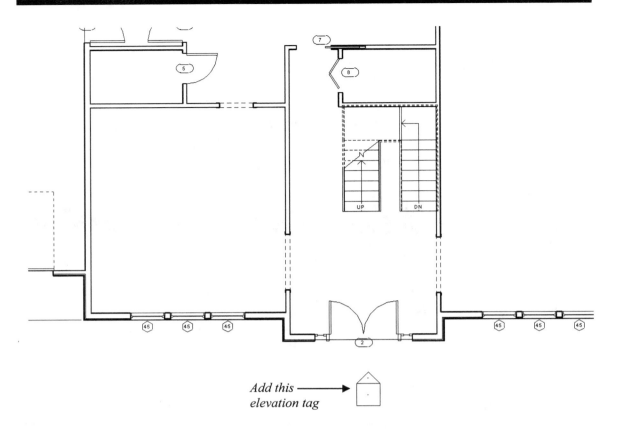

Add this ⟶
elevation tag

Figure 9-1.1 Added elevation tag

5. Right-click on the view label and select **Rename**.

6. Type: **South – Main Entry**.

The name should be fairly descriptive so you can tell where the elevation is just by the label. This will be essential on a large project that has several exterior elevations and even more interior elevations.

7. Double-click on **South – Main Entry** in the *Project Browser*.

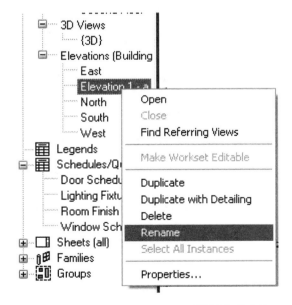

Figure 9-1.2
Renaming new view name

The elevation may not look correct right away. You will adjust this in the next step. Notice, though, that an elevation was created simply by placing an elevation tag in plan view.

8.   Switch back to the **First Floor** plan view.

Next you will study the options associated with the elevation tag. This, in part, controls what is seen in the elevation.

9.   The elevation tag has two parts: the pointing triangle and the square center. Each part will highlight as you move the cursor over it. **Select the square center part**.

You should now see the symbol shown on the right (Figure 9-1.3).

View direction boxes:
The checked box indicates which way the elevation tag is looking. You can check (or uncheck) the other boxes (Do NOT do that at this time). Each checked box adds another view name to the *Project Browser*.

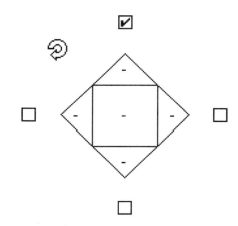

Rotation control:
Allows you to look perpendicular to an angled wall in plan, for example.

Figure 9-1.3 Selected elevation tag

Move elevation tag:
While selected, you simply drag the tag to move it.

10. Press the **ESC** key to unselect the elevation tag.

11. Select the "pointing" portion of the elevation tag.

Your elevation tag should look similar to Figure 9-1.4.

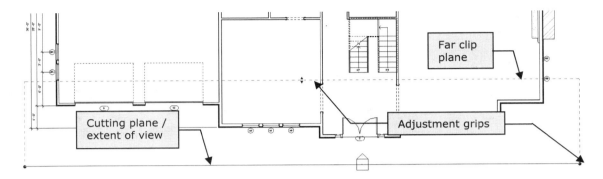

Far clip plane

Cutting plane / extent of view

Adjustment grips

Figure 9-1.4 Selected elevation tag

The elevation tag, as selected in Figure 9-1.4, has several features for controlling how the elevation looks. Here is a quick explanation:

- **Cutting plane/extent of view line:** This controls how much of the 3D model is elevated from left to right (i.e. the width of the elevation).
- **Far clip plane:** This controls how far into the 3D model the elevation view can see.
- **Adjustment grips:** You can drag this with the mouse to control the features mentioned above.

12. Right-click on the view label: **South – Main Entry** in the *Project Browser*, and then select **Properties**.

You have several options in the *Properties* window (Figure 9-1.5). Notice the three options with the check box next to them, these control the following:

- **Crop Region**: This crops the width and height of the view in elevation. *Adjusting the width of the cropping window in elevation also adjusts the "extent of view" control in plan view.*

- **Crop Region Visible**: This displays a rectangle in the elevation view indicating the extent of the cropping window (described above). *When selected in elevation view, the rectangle can be adjusted with the adjustment grips.*

- **Far Clip Plane Active**: If this is turned off, Revit will draw everything visible in the 3D model *(within the "extent of view").*

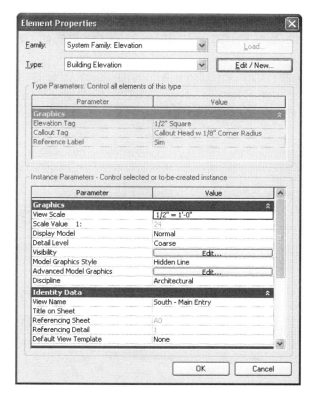

Figure 9-1.5 South – Main Entry Elev view properties

You will manipulate some of these controls next.

13. With the elevation tag still selected (as in Figure 9-1.4), drag (i.e. move) the "cutting plane/extent of view" line up into the main entry as shown in Figure 9-1.6.

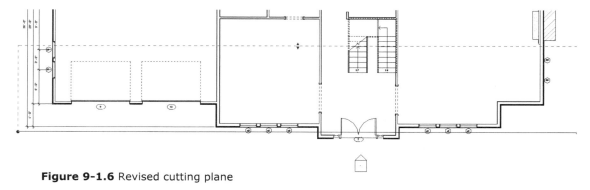

Figure 9-1.6 Revised cutting plane

14. Now switch to the *Elevation view*: **South – Main Entry**.

Your elevation should look similar to Figure 9-1.7. If required, click on the cropping window and resize it to match this image.

The main entry wall and roof are now displayed in section because of the location of the "cutting plane" line in plan.

Figure 9-1.7 Elevation with cutting plane through front entry

Notice that the roof is not fully visible. This is not related to the cropping window shown in Figure 9-1.7. Rather, it is related to the "Far Clip Plane" set in the plan view.

15. Adjust the "Far Clip Plane" in **First Floor** plan view so that the entire roof shows in the **South – Main Entry** view.
    TIP: Click on the grip and drag the "far clip plane" north until it is past the ridge line (which is at the middle of the building).

Now when you switch back to the South – Main Entry elevation view you should see all of the roof.

Next you will adjust the elevation tag to set up a detail elevation for the main entry area.

16. In the **First Floor** plan view, adjust the elevation tag to show only the main entry wall (Figure 9-1.8).
    - The Cutting Plane/Extent of View line is moved south so it is outside of the building footprint.
    - Use the left and right grips to shorten the same line.

17. Switch to **South – Main Entry** view to see the "detail" elevation you are setting up (Figure 9-1.9).

18. Make sure the **View Scale** is set to **1/2" = 1'-0"** on the View Control bar at the bottom of the screen.

Notice how the Level tags all got smaller, Undo and try it again if you missed it. You did notice the Level tags were added, right?

19. Adjust the South – Main Entry view's **Properties** to turn off the crop window's visibility.

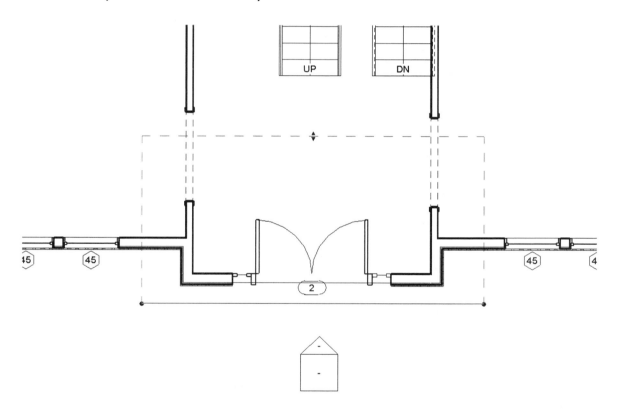

Figure 9-1.8 Main Entry wall detail elevation

Once you modify the view constraints in plan view (Fig. 9-1.8) and you switch to the new elevation view, you can click on the "Crop Window" and use the grips to make the view wider. You might want to do this to make sure the roof overhang shows.

When you drag the *Crop Window* to the right, the *Level* tags move so they don't overlap the elevation. This is another example of Revit taking the busy work out of designing a residence!

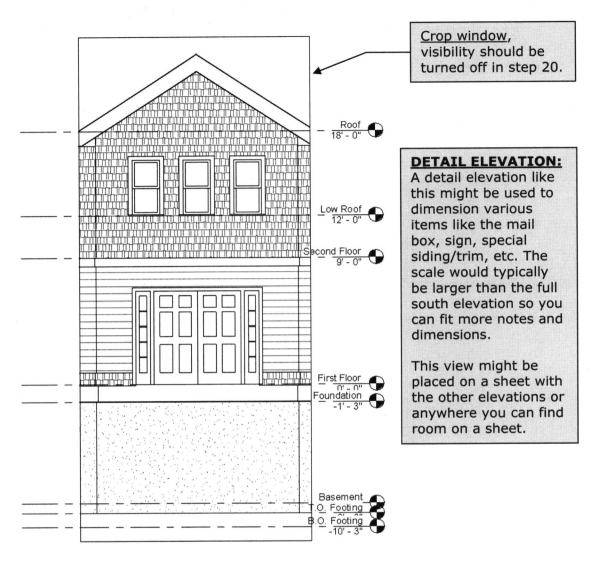

Crop window, visibility should be turned off in step 20.

Roof
18' - 0"

Low Roof
12' - 0"

Second Floor
9' - 0"

First Floor
0' - 0"

Foundation
-1' - 3"

Basement

T.O. Footing

B.O. Footing
-10' - 3"

DETAIL ELEVATION:
A detail elevation like this might be used to dimension various items like the mail box, sign, special siding/trim, etc. The scale would typically be larger than the full south elevation so you can fit more notes and dimensions.

This view might be placed on a sheet with the other elevations or anywhere you can find room on a sheet.

Figure 9-1.9 Main Entry wall detail elevation; View Scale is set to ¼" = 1'-0" so the level labels would be legible in this book.

20. **Save** your project as **ex9-1.rvt**.

Exercise 9-2:
Modifying the Project Model: Exterior Elevations

The purpose of this exercise is to demonstrate that changes can be made anywhere and all other drawings are automatically updated.

Modify an exterior elevation:

1.   Open ex9-1.rvt and **Save As ex9-2**.

2.   Open the **West** exterior elevation view.

3.   Using the Ctrl key, select the two double-hung windows on the right (i.e. two of the four windows in the garage wall).

You will delete the two selected windows and move the remaining two windows to the center of the wall.

4.   Press the **Delete** key to erase the selected windows

5.   Select the remaining two windows and use the **Move** tool to <u>approximately</u> center them on the wall / roof ridge, keeping the vertical alignment unchanged. (Figure 9-2.1)

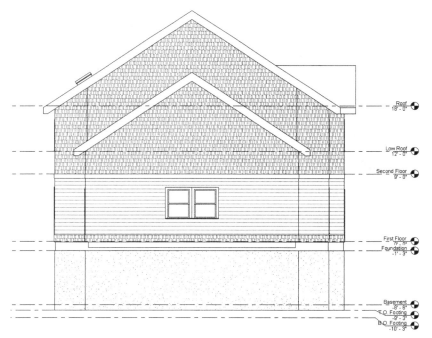

Figure 9-2.1 Modifying windows on the West elevation

Now you will switch to the first floor plan view to see your changes.

6. Switch to the **First Floor** plan view and zoom in on the Garage area. (Figure 9-2.2)

Notice how the windows in plan have changed to match the modifications you just made to the exterior elevations? This only makes sense seeing as both the floor plan and the exterior elevation are projected 2D views of the same 3D model. Both views are directly manipulating the 3D model.

Also notice that even the dimensions adjusted! Revit is all about reducing drafting time and maximizing design time. Step 5 had you approximately locate the windows, now you can select the windows in plan and type the exact dimension you want (12'-3" centers it on the building).

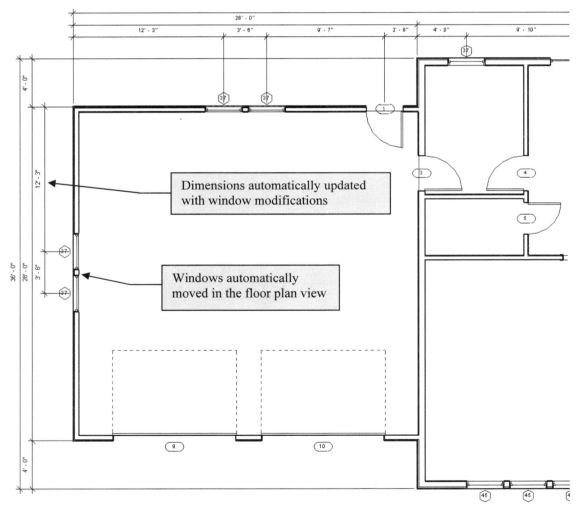

Figure 9-2.2 Changes automatically made to the floor plan

Next, you will insert a window in elevation. This will demonstrate, first, that you can actually add a window in elevation not just plan view, and second, that the other views are automatically updated.

7.   Switch to the **North** exterior elevation view.

8.   Select the **Window** tool.

Notice, with the window selected for placement, you have the usual dimensions helping you accurately place the window. As you move the window around you should see a dashed horizontal green line indicating the default sill height.

9.   From the *Type Selector*, choose: **Double Hung with Trim: 36" x 48"**.

10.   Place a window as shown in **Figure 9-2.3**; make sure the bottom of the window "snaps" to the green sill line.
      TIP: The window should be about centered between the existing windows.

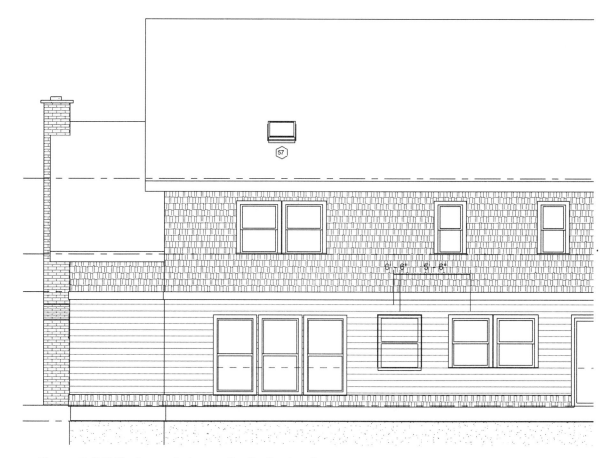

Figure 9-2.3 Placing a window on the North elevation

If you laid out the interior walls as described in Lesson 5, you should get a warning message when inserting the window. This is because the interior wall between the kitchen and the living room conflicts with the exterior window. Revit is smart enough to see that conflict and bring it to your attention. You can ignore this error (some you cannot), because it is possible, in the real-world, to build a wall up to the center of a window.

11. Click the "X" in the upper-right corner to ignore the wall/ window conflict warning. (Figure 9-2.4)

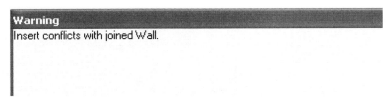

Figure 9-2.4 Conflict warning

12. Switch to the **First Floor** plan view to see the problem. (Figure 9-2.5)

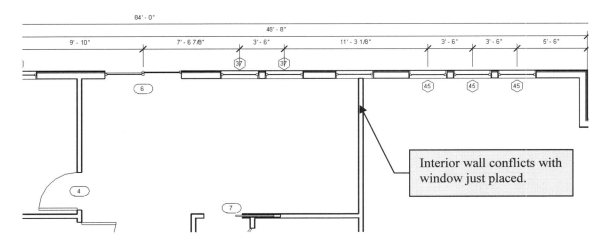

Figure 9-2.5 First Floor – kitchen area

13. While in plan view, select the new window and **Delete** it.

Why did the dimensions update in Figure 9-9.2 and not in Figure 9-2.5? The first had dimensions associated with the windows and the later did not. When you dimension something in Revit, the witness lines "remember" what they are dimensions to; and move with that element.

Adding shutters to the south elevation:

Next you will add shutters to the south elevation. You will need to load a family from the online Revit Content website.

14. Switch back to the **North** elevation view and notice that the new window is gone (per your plan change).

15. Select **Component** and then **Load** from the *Options Bar*.

16. Click the **Web Library** button.

17. Browse to **Revit Building 8 Subscription Library** → **Misc. Architectural**.

18. Save **Shutter (high-res).rfa** to your hard drive and then *Load* it. **TIP**: *Refer to previous lessons for more instruction.*

19. With the **Component** tool active and **Shutter (high-res): 18" x 48"** selected in the *Type Selector*, place a shutter in elevation as shown in Figure 9-2.6 (upper-left window on North Elevation). **TIP**: *If the shutter does not land in the correct spot, select it and use the arrow keys (on the keyboard) to nudge it into place.*

20. Use the same technique to place the remaining shutters shown in **Figure 9-2.7**.

21. Add shutters to the south elevation per **Figure 9-2.8**.

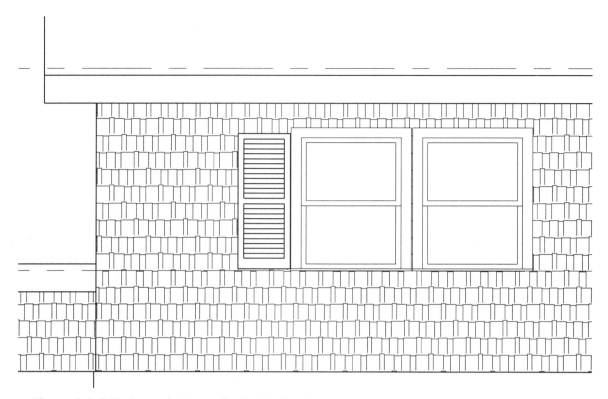

Figure 9-2.6 Placing a shutter on the North elevation

The shutter element has been setup so that it can only be placed on a wall. If you move the cursor over a door or window you get the (⊘) symbol – meaning it cannot be placed there.

Figure 9-2.7 Shutters (8) added to the North elevation

Figure 9-2.8 Shutters (4) added to the South elevation

22. **Save** your Project as ex9-2.rvt.

REMINDER: ENTERING DIMENSIONS IN REVIT

```
0 48|
```

As your experience with Revit grows, you will want to learn some of the shortcuts to using the program. One of those shortcuts is how you enter dimensions when drawing. You probably already know, maybe by accident, that if you enter only one number (e.g., 48) and press enter, Revit interprets that number to be feet (e.g., 48'-0"). So, if you want to enter 48", you may be typing 0'-48" or 48". Both work, but having to press the Shift key to get the inch symbol takes a little longer.

Here are some options for entering dimensions:

0 48	*Revit reads this as 48" (zero space forty-eight)*
48	*Revit reads this as 48'-0"*
5.5	*Revit reads this as 5'-6"*
0 5.5	*Revit reads this as 5½"*
2 0 1/4	*Revit reads this as 2'-0¼" (two space zero space fraction)*

Exercise 9-3:
Creating & Viewing Parametric Interior Elevations

Creating interior elevations is very much like exterior elevations. In fact, you use the same tool. The main difference is that you are placing the elevation tag inside the building, rather than on the exterior.

Adding an interior elevation tag:

1. Open project ex9-2.rvt and **Save As ex9-3.rvt**.

2. Switch to the **First Floor** floor plan view, if necessary.

3. Select the **Elevation** tool. 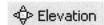 Elevation

In the next step you will place an elevation tag. Before clicking to place the tag, try moving it around to see how Revit automatically turns the tag to point at the closest wall.

4. Place an elevation tag, looking east (i.e. to the right), in the living room area. (Figure 9-3.1)

Remember, the first thing you should do after adding an elevation tag is to give it an appropriate name in the *Project Browser* list.

5. Change the name of the elevation to **Living Room - East**.

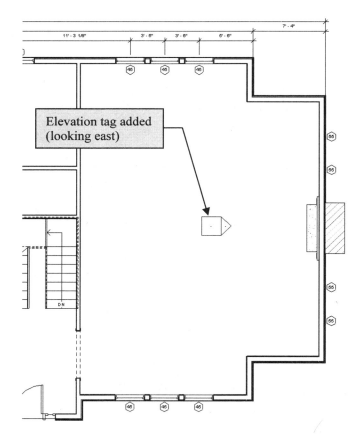

Figure 9-3.1 First floor: Elev tag added in Living Room

6. Switch to the **Living Room – East** view.
    TIP: Try double-clicking on the elevation tag (the triangular portion).

Initially, your elevation should look something like Figure 9-3.2. You will adjust this view next. *Notice how Revit automatically controls the lineweights of things in section vs. things in elevations.*

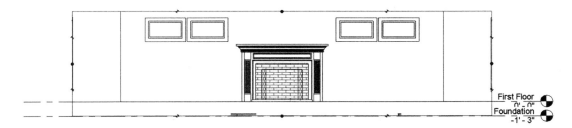

Figure 9-3.2 Living Room - East – initial view

7. Switch back to the **First Floor** plan view.

8. Pick the "pointing" portion of the elevation tag (triangle), so you see the view options. (Figure 9-3.3)

You should compare the two drawings on this page (Figures 9-3.2 & 9-3.3) to see how the control lines in the plan view dictate what is generated/ visible in the elevation view, for both width and depth.

Revit automatically found the left and right walls, the floor and ceiling in the elevation view.

Notice the Far Clip Plane is also acceptable. If you cannot see the fireplace in the elevation view; that means the Far Clip Plane needs to be moved further to the right.

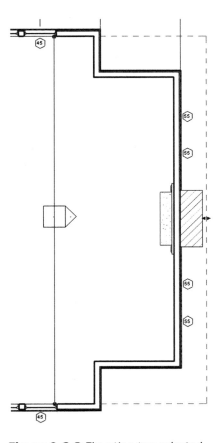

Figure 9-3.3 Elevation tag selected

> **FYI:** The elevation tags are used to reference the sheet and drawing number so the client or contractor can find the desired elevation quickly while looking at the floor plans. This will be covered in a later lesson. It is interesting to know, however, that Revit automatically does this (fills in the elevation tag) when the elevation is placed on a sheet, and will update it if the elevation is moved.

9. Switch back to the **Living Room - East** view.

Just to try adjusting the Crop Window to see various results, you will extend the top upward to see the second floor.

10. Select the *Cropping Window* and drag the top middle grip upward, to increase the view size vertically. (Figure 9-3.4)

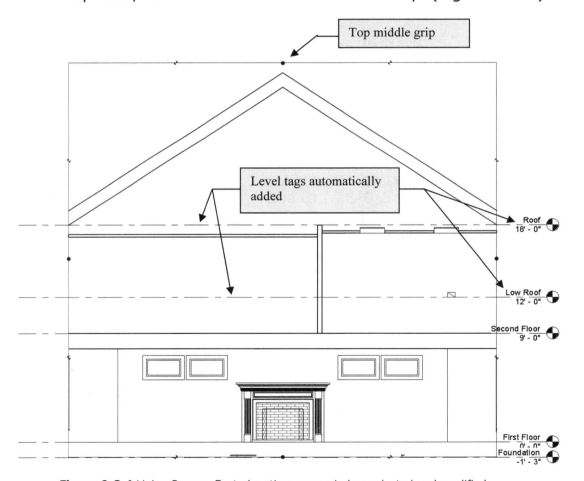

Figure 9-3.4 Living Room - East elevation: crop window selected and modified

Here you are getting a sneak peak ahead to the "sections" portion of this book. Notice that additional Level tags where added automatically as the view grew to include more vertical information (i.e. second floor, roof).

11. Now select **Undo** to return to the previous view. (Figure 9-3.2)

As you probably noticed, the interior elevation shows the Foundation Level tag below the First Floor Level tag. The Foundation Level tag is not necessary in this view, so you will remove it from this view.

IMPORTANT: You cannot simply delete the Level tag because it will actually delete it from the entire project; not only the Foundation Level tags but the Foundation Floor plan view, etc.

To remove a Level tag, you select it and tell Revit to hide it in the current view. You will do this next.

12. Click on the horizontal line portion of the Level tag.

13. Right-click and select **Hide Annotation in View** from the pop-up menu.

14. Adjust the bottom of the Crop Window to align with the First Floor level (horizontal line).

15. On the **View Control** bar, set the *View Scale* to **1/4"=1'-0"**.

Your elevation should look like Figure 9-3.5.

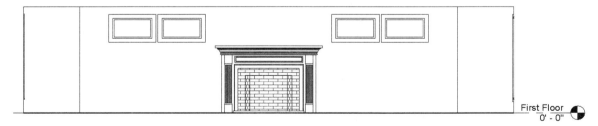

Figure 9-3.5 Living Room - East elevation

You can leave the crop window on to help define the perimeter of the elevation. You can also turn it off. However, some lines that are directly under the crop window might disappear. You could use the Line tool to manually define the perimeter (using a heavy line).

Adding annotation:

Now you will add a note and dimension to the interior elevation view.

16. Add the (2) dimensions and (1) text w/ leader shown in Figure 9-3.6.

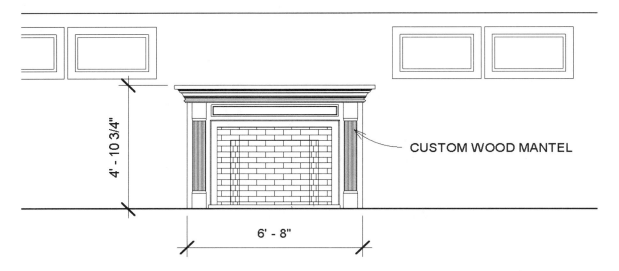

Figure 9-3.6 Living Room - East elevation: annotation added

17. On the **View Control** bar, set the *View Scale* to **1/2"=1'-0"**.

Notice the level tag, text and dimensions automatically resized to match the new scale. When space permits, most interior elevations are 1/2" = 1'-0".

18. Save your project as **ex9-3.rvt**.

Exercise 9-4:
Modifying the Project Model: Interior Elevations

This short exercise, similar to exercise 9-2, will look at an example of Revit's parametric change engine. All drawings are generated from one 3D model.

Modify the interior elevations:

1. Open ex9-3.rvt and **Save As ex9-4.rvt**.

2. Open the **Living Room - East** elevation view.

You will move the windows and add an exterior door.

3. Select both of the windows on right; use the Ctrl key to select multiple objects at one time.

4. Use the **Move** tool to move the windows 6" to the right (south).

5. Repeat the previous steps to Move the other two windows 6" towards the left (north).

6. In the *Living Room - East* elevation view, use the **Door** tool to place a **Single Entry 3: 32" x 84"** door in the wall to the far left (north). See Figure 9-4.1.

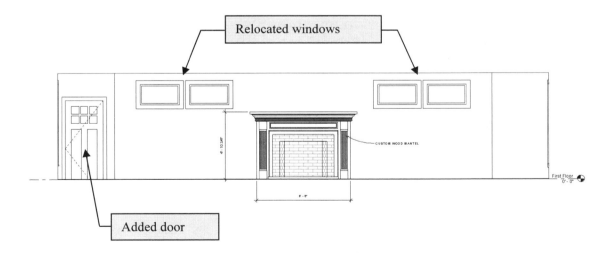

Figure 9-4.1 Living Room – East – modified

You moved the windows out from the mantel so they provided a little more space between them. Revit allows you to design in all views!

You should notice, in Figure 9-4.1, that the door bottom is below the floor level; this is a problem. You will fix it next.

7. Select the new door in the *Living Room – East* view.

8. In *Properties*, change the **Level** from *Foundation* to **First Floor**.

The interior elevation should now look correct (Figure 9-4.2).

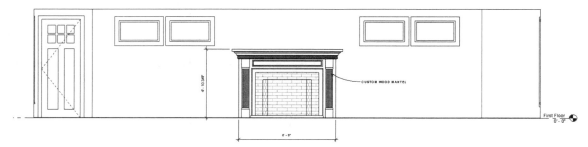

Figure 9-4.2 Living Room - East elevation: door height adjusted

Notice, back in *Properties* for the new door, a setting called "sill height". This setting allows you to move the door vertically, relative to the specified *Level*. For example, you might have a second floor door that opens onto a low roof; but the door needs to be 1'-0" above the second floor for this door to work correctly (roof insulation, flashing, etc.). You simply select the door and adjust its "sill height" to 1'-0" (with the Level set to second floor) via *Properties*.

Now it's time to see the effects to the plan views.

9. Switch to the **First Floor** plan view. (Figure 9-4.3)

When placing a door in elevation, you may have to switch to plan view to verify the door swing is the way you want it; you cannot control the door swing in elevation.

In elevation, you can adjust many things this way. Some examples are: ceiling height, interior and exterior windows, wall locations (perpendicular to the current view), etc.

Notice that the door does not have a door tag like the other doors in the plan view. This is because the door was not placed in this view. You will learn how to add this tag later.

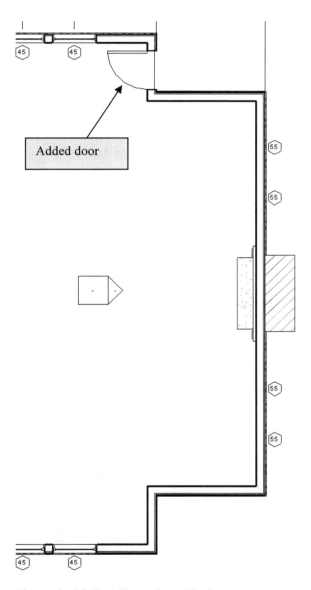

Figure 9-4.3 First Floor plan with changes

10. Save your project as **ex9-4**.

Exercise 9-5:
Design Options

This exercise will look at a relatively new feature called *Design Options*. This feature allows you to present two or more options for a portion of your design without having to save a copy of your project and end up having to maintain more than one project until the preferred design option is selected.

Design Options Overview:

A Revit project can have several design option studies at any given time; you might have an (A) entry porch options study, (B) a kitchen options study and a (C) master bedroom furniture layout options study in a project. Each of these studies can have several design options associated with them. For example, the entry porch study might have three options: 1. flat roof, 2. shed roof, and 3. shed roof with a dormer type embellishment.

A particular study of an area within your project is called a *Design Option Set*, and the different designs associated with a *Design Options Set* are each called an *Option*. Both the *Design Options Set* and the *Options* can be named. (See the examples below.)

Design Option Set

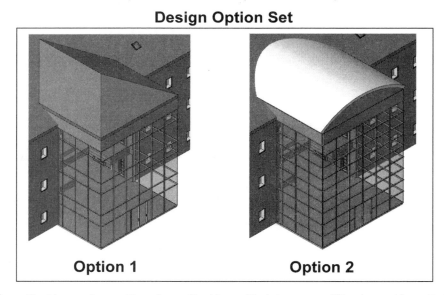

Option 1   **Option 2**

One of the *Options* in a *Design Option Set* is specified as the *Primary* option (the others are called *Secondary Options*); this is the option that is shown by default in all new and existing views. However, you can adjust the *Visibility* of a *View* to show a different option. Typically you would duplicate a *View*, adjust the *Visibility*, and *Rename* it to have each option at your finger-tips.

When the preferred design is selected, by you and/or the owner, you set that *Option* to *Primary*. Finally, you select a tool called "*Accept Primary*" which deletes the *Secondary Options* and the *Design Option Set*, leaving the *Primary Option* as part of the main building model.

The following image (Figure 9-5.1) is an overview of the *Design Options* dialog box. (See step 2 on page 9-27, to access this dialog.)

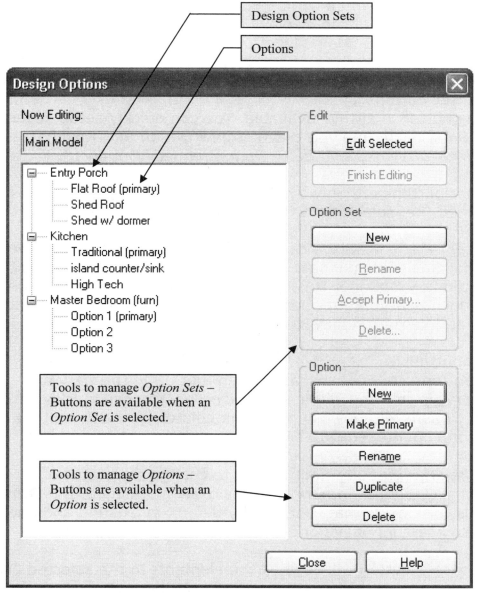

Figure 9-5.1 Design Options Dialog

Notes about the Design Options dialog box (next page):

Edit buttons:
You can edit a *Design Option* by selecting an *Option* (in the window area on the left) and then clicking the "*Edit Selected Option*" button. Next you add, move and delete elements in that Design Option.

When finished editing a *Design Option*, you reopen the *Design Options* dialog and click the "*Finish Editing Option*" button.

If you are currently in an Option Editing mode, the *Now Editing* area in the *Design Options* dialog displays the *Option* name being modified; otherwise it displays "Main Model".

Option Set buttons:
The *New* button is always available. You can quickly setup several *Option Sets*. Each time you create a new *Option Set*, Revit automatically creates a *Primary Option* named "Option 1".

The other buttons are only available when an Option Set label is highlighted (i.e. selected) in the window list on the left.

The *Accept Primary* button causes the Primary option of the selected Set to become a normal part of the building model and deletes the Set and Secondary Options. This is a way of "cleaning house", by getting rid of unnecessary information which helps to better manage the project and keeps the file size down.

Option buttons:
These buttons are only available when an *Option* (primary or secondary) is selected within an *Option Set*. You can quickly setup several *Options* without having to immediately add any content (i.e. walls, components, etc.) to them.

The *Make Primary* button allows you to change the status of a *Secondary Option* to *Primary*. As previously mentioned, the *Primary Option* is the *Option* that is shown by default in existing and new views. You can only have one *Primary* option in an *Option Set*.

The *Duplicate* button will copy all the elements in the selected Option into a new Option (this makes the file larger because you are technically adding additional content to the project). You can them use the copied elements (e.g. walls, furniture, etc.) as a starting point for the next design option. This is handy if the various options are similar.

Now you will put this knowledge to use!

Setting up a Design Option Set:
In this exercise you will create two *Design Option Sets*; one for the Main Entry Roof and another for the windows above the Entry Door. You will create an alternate roof and window design for your project.

You could just create one *Design Option Set* and have two design options total. However, by placing the roof options in one *Option Set* and the windows in another, you actually get four design options total; you can mix and match the window and roof options.

Setting up Design Options in your project:

First you will setup the Option Sets and Options.

1. Open ex9-4.rvt and **Save As ex9-5.rvt**.

2. Select *Design Options* → **Design Options...** from the *Tools* pull-down menu.

You are now in the *Design Options* dialog box (Figure 9-5.2), unless you have modified your template file to have *Option Sets*, your dialog will look like this one.

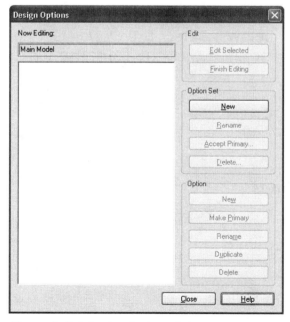

Figure 9-5.2 Design Options dialog; initial view

3. In the *Option Set* area click **New**.

Notice an *Option Set* named <u>Option Set 1</u> has been created. Revit also automatically created the *Primary Option* named <u>Option 1</u>. (Figure 9-5.3). Next you will rename the Option Set to something that is easier to recognize.

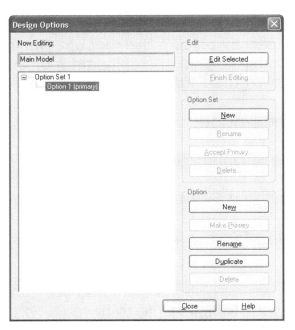

Figure 9-5.3
Design Options dialog; new option set created

4. (See warning below.) Select the *Option Set* currently named <u>Option Set 1</u> and then click the **Rename** button in the *Option Set* area.

   WARNING! be sure you are not renaming the *Option*, but rather, the *Option Set.*

5. In the *Rename* dialog type: **Entry Windows**. (Figure 9-5.4).

6. Click **OK** to rename.

Giving the *Option Set* a name that is easy to recognize helps managing the various options later, especially if you have several.

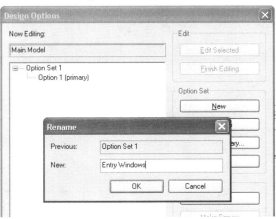

Figure 9-5.4
Rename Option Set dialog

Next you will create a *Secondary Option* for the *Entry Windows* Option *Set*.

7. With the *Entry Windows Option Set* selected (or any option in that set), click **New** in the *Option* area. (Figure 9-5.5)

Notice a secondary Option was created and automatically named Option 2. If you have descriptive names for the options in a set, you should apply them. In this example you can leave them as is.

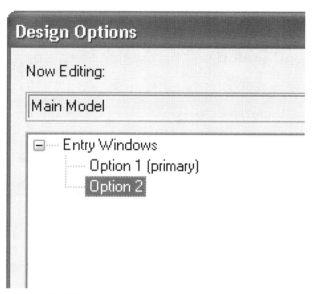

Figure 9-5.5
Design Options dialog; secondary option created

8. Create an *Option Set* for the roof (Figure 9-5.6):
   a. Name the set: **Entry Roof**
   b. Create one secondary option

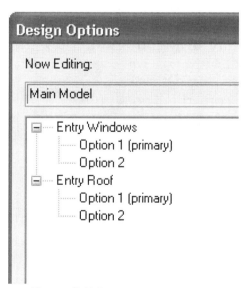

Figure 9-5.6
Design Options dialog

The basic thinking with the *Design Options* feature is that you setup the *Option Set* and *Option*s and then start drawing the elements related to the current *Option*. So, you create the *Options* and then click "Edit Selected Option" in the *Design Options* dialog, close the dialog, make the additions and modifications relative to that *Design Option*, and finally go back into the *Design Option* dialog and click the "Finish Editing Option" button.

However, in your case, you want to move content already drawn to *Option 1*. Revit has a feature that allows you to move content to a *Design Option Set*; which means the content gets copied to each *Option* in the *Set* you select. This option will work for the windows because the second option will be similar to the first one.

Entry Windows Design Option:

You are now ready to setup the different design options.

9. Switch to the **3D view,** adjust the view so you are looking at the main entry (South elevation).

Unfortunately, you cannot just select the windows and add them to a Design Option Set because they are host elements that require a wall to be inserted in. Therefore, you will also add the entry walls and door to the Design Option Set.

10. Select the three walls at the main entry, three windows and main entry door.

11. From the *Tools* pull-down menu, select *Design Options* → **Add to Design Option Set**.

12. Select *Entry Windows* from the dialog drop-down and then click **OK**. (Figure 9-5.7)

Figure 9-5.7
Design Options dialog

The selected items are now in both _Option 1_ and _Option 2_ under the _Option Set: Entry Windows_ (because both options were selected in the previous dialog, see Figure 9-5.7).

From this point forward you can only modify the entry windows/door/walls when in _Option 1_ or _Option 2_ "edit mode" (in which case the tables are turned and you cannot edit the main building model; this is because "exclude options" is selected on the _Options Bar_).

13. In the **Design Options** dialog, select **Entry Windows:Option 2** (i.e. Option Set: _Entry Windows_; Option: _Option 2_).

14. Click the **Edit Selected Option** button. (Figure 9-5.1)

15. Click **OK** to close the _Design Options_ dialog.

Now you should notice that the main building model is slightly grey and not editable (it is not editable because "active option only" is selected on the _Options Bar_). You will change the windows per Figure 9-5.8).

16. Zoom in on the main entry, particularly the second floor window area in your 3D view.

17. Select the three windows above the main entry door change their type to **Double Hung with Trim: 16" x 72"** from the _Type Selector._

18. Make sure the window head (i.e. top) is at **7'-0"** via Properties for the selected windows.

Figure 9-5.8
3D View: second floor windows modified

19. Open the *Design Options* dialog and click the **Finish Editing Option** button.

20. Click **OK** to close the dialog box.

It now appears like all your changes disappeared, right? Well, if you recall from the introduction to this exercise, the *Primary Option* is displayed by default for all new and existing views. So when you finished editing *Entry Windows:Option 2* the *3D View* switched back to the Primary Option (which is currently set to Option 1).

Next you will create a new view and adjust its *Visibility* to display *Option 2* of the *Entry Windows* Options Set.

First you will create a duplicate copy of the 3D view.

21. In the Project Browser, under 3D Views, right-click on the {3D} label.

22. Select **Duplicate** from the pop-up menu.

You now have a copy of the 3D view named *Copy of {3D}*.

23. **Rename** the new view to **Entry Windows Option 2**.

24. Switch to your new view. (Revit may have switched it for you.)

25. From the *View* pull-down menu, select **Visibility/Graphics**.

26. Click on the **Design Options** tab at the top of the dialog.

27. Change the *Design Option* parameter for *Entry Windows* to *Option 2*. (Figure 9-5.9)

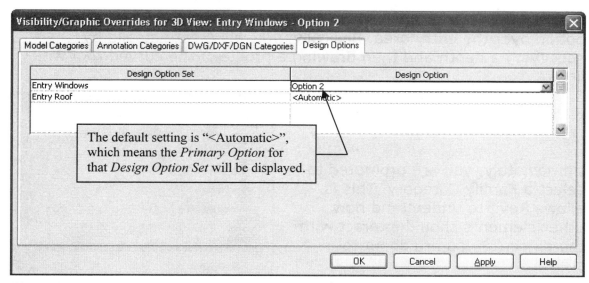

Figure 9-5.9
Visibility/Graphic Overrides dialog: modified Entry Windows design option visibility

28. Click **OK** to close the dialog.

Now, with the new 3D view and the *Entry Windows:Option 2* view, you can quickly switch between design options. Both views could be placed on the same sheet and printed out for a design critique.

Entry Roof Design Option:
Now you will add an option for a roof over the entry door.

29. Open the **Design Options...** dialog from the *Tools* menu.

30. Select **Option 2** under *Entry Roof Option Set*.

31. Click the **Edit Selected Option** button.

32. Click **OK** to close the *Design Options* dialog.

You will add a roof shortly.

33. Switch to the **South** elevation view.

34. Zoom in to the area below the entry windows.

Next you will create an in-place Family to represent a curved roof option over the entry area. Basically, you will create a solid by specifying a depth and then drawing a profile of the curved roof with lines.

35. Click **Create...** on the *Modeling* Tab.

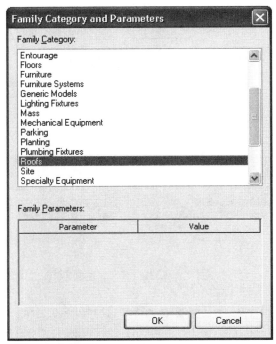

Immediately, you are prompted to select a Family Category. This is allows Revit to understand how other elements should interact with the object(s) you are about to create.

36. Select **Roofs** from the *Family Category* list. (Figure 9-5.10)

37. Click **OK**.

Now you are prompted to provide a name for the new *Family*.

Figure 9-5.10
Family Category and Parameters dialog

38. For the *Family* name, type **Entry Roof**. (Figure 9-5.11)

Figure 9-5.11
Family name prompt

You are now in a mode where you draw the Entry Roof. Notice that the *Design Bar* has changed to just have a *Family* tab; which has all the tools available to create a *Family*. You are continuously in the *Family Edit* mode until you select *Finish Family* or *Quit Family* from the *Design Bar*.

39. Click **Solid Form** on the *Design Bar*.

You are now prompted to select the type (or form) of solid you wish to create.

40. Select **Solid Extrude** on the popup menu. (Figure 9-5.12)

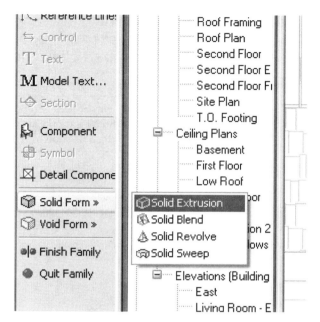

Figure 9-5.12 Solid Form popup menu; select Solid Extrude

Finally, you are prompted to select a plane in which to start drawing the profile of the solid to be extruded. Even though the view is a 2D representation of a 3D model, Revit needs to know where you want the 3D Solid created. You will select the entry wall as a reference surface to establish a working plane.

41. Select **Pick a Plane** and click **OK**. (Figure 9-5.13)

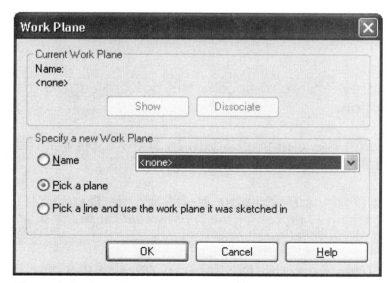

Figure 9-5.13 Work Plane dialog; select Pick a Plane

42. Move the cursor near the edge of the entry wall (*TIP: click the Tab key until a dashed line appears around the perimeter of the wall if required*), and then click the mouse to select. (Figure 9-5.14)

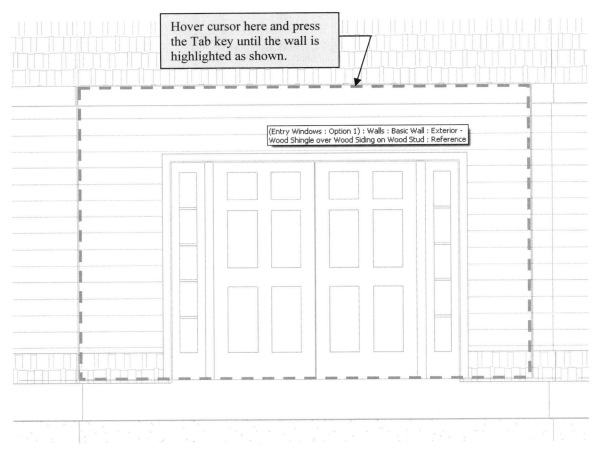

Hover cursor here and press the Tab key until the wall is highlighted as shown.

(Entry Windows : Option 1) : Walls : Basic Wall : Exterior - Wood Shingle over Wood Siding on Wood Stud : Reference

Figure 9-5.14 South elevation; select wall to establish work plane

Next you will draw an arc to specify the bottom edge of the curved roof design option.

Notice the Design Bar changed again to show tools related to drawing an extruded solid (Figure 9-5.15).

43. On the *Options Bar*, enter **2'-0"** for the **Depth**. (Figure 9-5.15)
FYI: # negative number for the depth would cause the solid to project in, towards the building from the work plane.

44. Click **Lines** from the *Design Bar*. (Figure 9-5.15)

45. Click the **Arc** icon from the *Options Bar* (the one on the left; arc passing through three points).

46. Pick the three points shown in Figure 9-5.15 to draw the arc; the angle 71.656° is not critical, get as close as possible; the ends of the arc are approx. straight up from the edge of the entry door below.

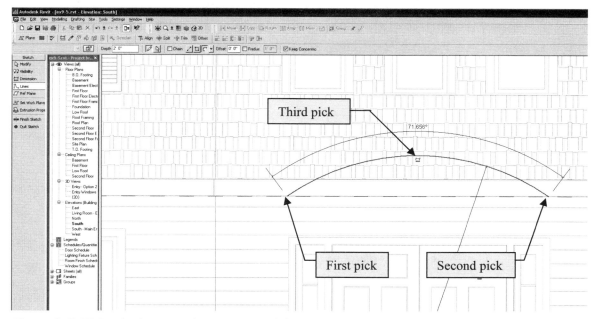

Figure 9-5.15 South elevation; drawing arc to define roof

Now you will draw another arc 7¼" above the previous one.

47. On the *Options Bar*, enter **0'-7.25"** for the *Offset*.

48. Pick the same three points shown in Figure 9-5.15.

Notice that an arc is drawn offset 7¼" from the points you picked. If you pick the first two points in the other direction, the arc would be offset in the other direction (downward in this case).

Next you will draw two short lines to connect the endpoints of the two arcs. This will create a closed area, which is required before finishing the sketch. Think of it this way: you need to completely specify at least two dimensions before Revit and extrude to create the third.

49. Click the "straight" line icon on the *Options Bar*; this will switch you from drawing arcs back to drawing straight line segments. Set the *Offset* back to **0"**.

50. Zoom in and sketch a short line on each end of the arc as shown in Figure 9-5.16.

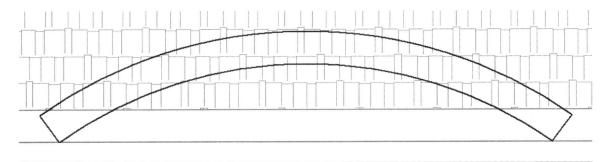

Figure 9-5.16 South elevation; two arcs and two short lines define roof profile

51. Click **Finish Sketch** from the *Design Bar*.
    TIP: *If you get any warnings, it may be because one or more of the profiles corners do not create a perfect intersection; zoom in to see and use the trim tool to close the corners.*

You are still in the "Create Family" mode. Before you finish you will apply a material to the roof element.

52. Click the roof to select it.

53. Click the **Properties** icon on the *Options Bar*.

54. Click in the *Material* value field and then click the down-**arrow icon** that appears.

The Material dialog opens.

55. Select **Metal –Roofing** from the list of predefined *Materials* and then click OK.
    FYI: *Notice the Accurneder material is set to "Stain, Plain" for Metal – roofing.*

56. Click **OK** to close the *Properties* dialog.

You are now ready to finish the Family.

57. From the *Design Bar* click **Finish Family**.

Next you will move the roof down a little. At this point, the only "mode" you are in is for editing the *Entry Roof: Option 2*.

58. Click on the roof to select it.

59. Use the arrow keys on the keyboard to nudge the roof element down until it looks similar to Figure 9-5.17.

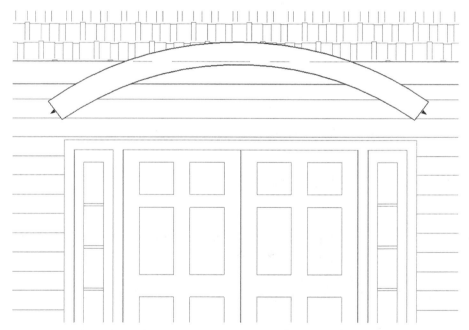

Figure 9-5.17 South elevation; roof element moved to a lower position

Notice the arrow-grips on each end of the roof element when it is selected? Clicking one of these grips allows you to extend that solid perpendicular to the arrow. Try it; just Undo when you are done experimenting. It works better on rectilinear objects.

You are now ready to finish editing the current design option for the time being.

60. In the *Design Options* dialog click **Finish Editing Option**.

61. Click **OK** to close the dialog.

As before, the Option you were just working on was not the *Primary Option* in the *Entry Roof Design Set* so the current view diverted back to *Entry Roof:Option 1* (which is the Primary view).

You will create a 3D view that has *Option 2* set to be visible for both the *Entry Window* Option Set and the *Entry Roof* Option Set.

62. Right-click on the **Default 3D view** and click **Duplicate**.

63. Rename the duplicated view to **Entry – Option 2**.

64. Switch to the new view (Entry – Option 2).

65. Click **Visibility/Graphics...** from the *View* menu.

66. On the *Design Options* tab, set both *Option Sets* to **Option 2** in the *Design Option* column.

67. Click **OK** to close the dialog.

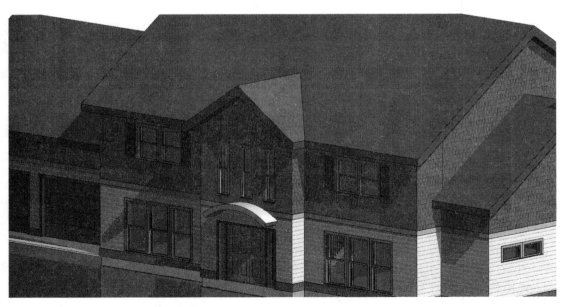

Figure 9-5.18 Entry – Option 2 view with shadows on

You can now see a 3D view of your new entry roof option. The previous image has shadows turned on. This can provide a very dramatic response with almost no extra effort. The shadows can be useful while designing; however, you will probably have them turned off most of the time just to maintain system performance.

To turn on Shadows, you simply click the shadows icon on the view control bar at the bottom of the screen, and then click Shadows On from the popup menu.

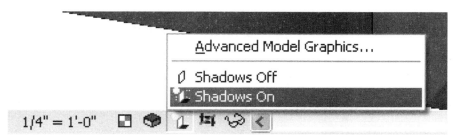

Because the new entry roof element is a roof to Revit, you can attach walls to it (meaning a wall top would conform to the underside of the curved roof.

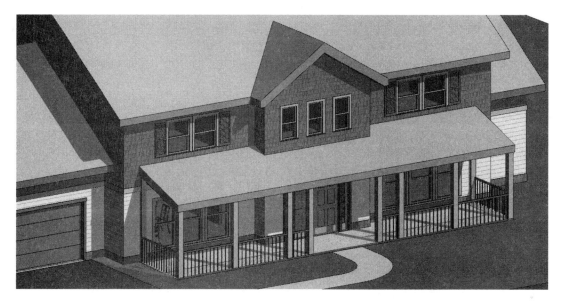

Another Entry roof option might be a more traditional shed type porch roof like the one shown above. You might try adding this roof to the Option 1 place holder. Simply sketch a roof by footprint, with only the south edge set to "define slope".

68. **Save** your project as **ex9-5.rvt**.
    TIP: The Design Options feature can also be used to manage alternates, where both the base bid and the alternate(s) need to be drawn.

Self-Exam:

The following questions can be used as a way to check your knowledge of this lesson. The answers can be found at the bottom of this page.

1. The plan is updated automatically when an elevation is modified, but not the other way around. (T/F)

2. You can use the Elevation tool to place both interior and exterior elevations. (T/F)

3. You can rename elevation views to better manage them. (T/F)

4. You have to resize the level tags and annotations after changing a view's scale. (T/F)

5. How do you enter 5 ½" without entering the foot or inch symbol?

   _____

Review Questions:

The following questions may be assigned by your instructor as a way to assess your knowledge of this section. Your instructor has the answers to the review questions.

1. The visibility of the crop window can be controlled. (T/F)

2. You have to manually adjust the lineweights in the elevations. (T/F)

3. As you move the cursor around the building, during placement, the elevation tag turns to point at the building. (T/F)

4. There is only one part of the elevation tag that can be selected. (T/F)

5. You cannot adjust the "extent of view" (width) using the crop window. (T/F)

6. What is the first thing you should do after placing an elevation tag?

   _____

7. Although they make the drawing look very interesting, using

   the _____ feature can cause Revit to run extremely slow.

8. With the elevation tag selected, you can use the _____ to adjust the tag orientation to look at an angled wall.

9. You need to adjust the _____ to see objects, in elevation, that are a distance back from the main elevation.

10. What feature allows you to develop different ideas? _____

Lesson 10
Residence: SECTIONS::

Sections are one of the main communication tools in a set of architectural drawings. They help the builder understand vertical relationships. Architectural sections can occasionally contradict other drawings, such as mechanical or structural drawings. One example is a beam shown on the section is smaller than what the structural drawings call for; this creates a problem in the field when the duct does not fit in the ceiling space. The ceiling gets lowered and/or the duct gets smaller, ultimately compromising the design to a certain degree.

Revit takes great steps toward eliminating these types of conflicts. Sections, like plans and elevations, are generated from the 3D model. So it is virtually impossible to have a conflict between the architectural drawings. The final step will be to get the engineers working on the same 3D model; this would eliminate conflicts and redundancy in drawing.

Exercise 10-1:
Specify section cutting plane in plan view

Similar to elevation tags, placing the reference tags in a plan view actually generates the section view. You will learn how to do this next.

Placing section tags:

1. Open ex9-5.rvt and **Save As ex10-1.rvt**.

2. Switch to **First Floor** plan view.

3. Select the **Section** tool from the **Basics** tab on the *Design Bar*.   ⬡ Section

4. Draw a **Section** tag as shown in Figure 10-1.1. Start on the left side in this case. Use the Move tool if needed to accurately adjust the section tag after insertion. The section should go through the stair landings. (Figure 10-1.1)

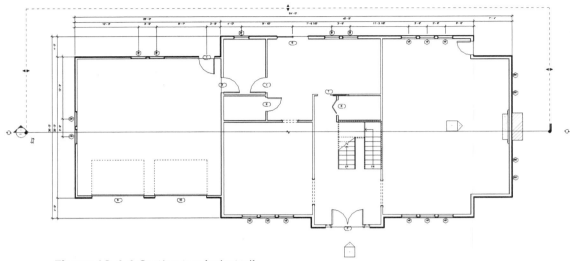

Figure 10-1.1 Section tag (selected)

You can see that the stairs in the *Far Clip Plane* location in the plan view are out past the building perimeter; this means you will see everything (Figure 10-1.1). Anything passing through the section line is in section; everything else you see is in elevation. Also, you can see the roof is shown in section exactly where the section line is shown in plan. Figure 10-1.2 also shows the *Crop Region*.

The *Section* tool is almost exactly like the *Elevation* tool.

Figure 10-1.2 Longitudinal view – updated

Notice, in the *Project Browser*, that a new category has been created: <u>Sections (building Section)</u>. If you expand this category you will see the new section view listed; it is called *Section 1*. You should always rename any new views right after creating them. This will help with navigation and when it comes time to placing views on sheets for printing.

5. Rename section view *Section 1* to **Longitudinal Section**.

6. Switch to the new section view, **Longitudinal Section.**
    TIP: *Double-click on the section tag's bubble to open the tags view.*

7. Change the view *Properties* so the **Detail Level** is set to **Medium**. *(Notice how the walls change to show more detail.)* You can set this on *the View Control Bar* at the bottom of the screen.

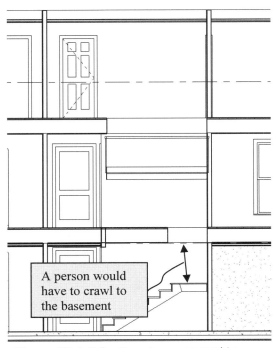

A person would have to crawl to the basement

Figure 10-1.3 Section view – zoomed in

8. Zoom in on the portion of the section through the stairs.

You should notice an added level of detail in the section view.

You may also notice a big problem; in Figure 10-1.3, the stair to the basement is obstructed by the first floor system. You will fix this soon.

Next you will add a cross sectional view.

9. Create a **Section** as shown in **Figure 10-1.4**. *TIP: You can use the control arrows to make the section look the other direction.*

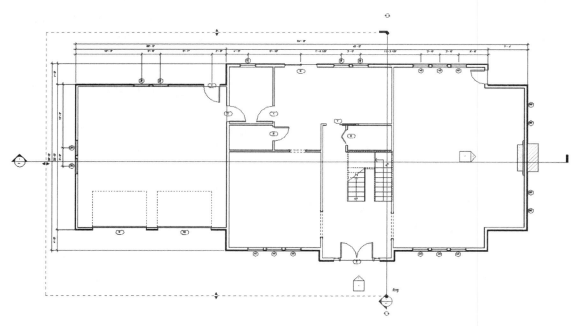

Figure 10-1.4 plan view; Section tag (selected)

10. Rename the new section view to **Cross Section 1** in the *Project Browser*.

11. Switch to the **Cross Section 1** view.

12. Set the *Detail Level* to **Medium** and turn off the **Crop Region** visibility in the *View Properties*. (Figure 10-1.5)

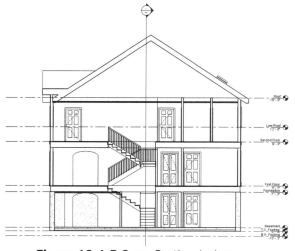

Revit automatically displays lines heavier for objects that are in section than for objects beyond the cutting plane and shown in elevation.

Also, with the Detail Level set to medium, the walls and floors are hatched to represent the material in section.

Figure 10-1.5 Cross Section 1 view

Notice that the *Longitudinal Section* tag is automatically displayed in the *Cross Section 1* view. If you switch to the *Longitudinal Section* view you will see the *Cross Section 1* tag. Keeping with Revit's philosophy of change anything anywhere, you can select the section tag in the other section view and adjust its various properties, like the *Far Clip Plane*.

FYI: In any view that has a Section Tag in it, you can double-click on the round reference bubble to quickly switch to that section view.

Modifying the first floor stair opening:

Next you will fix the problem with the stair opening in the first floor system. Basically, you will switch to the first floor plan view and select the floor. Then you will click *Edit* from the *Options Bar* and modify the sketch so the stair opening is larger.

13. Switch to the **First Floor** plan view.

14. Select the floor (the Tab key is required to highlight the floor element for selection).

15. Click **Edit** on the *Options Bar*.

16. Use the line and edit tools to modify the sketch of the floor opening as shown in Figure 10-1.6.

17. Click **Finish Sketch**.

18. Switch back to the **Longitudinal Section** view.

Notice the stair no longer has a problem (Figure 10-1.7).

19. **Save** your project as **ex10-1.rvt**.

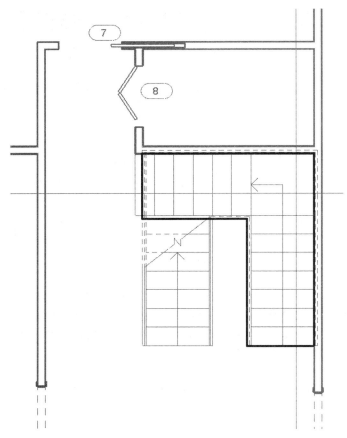

Figure 10-1.6 Plan view – Floor, sketch edit

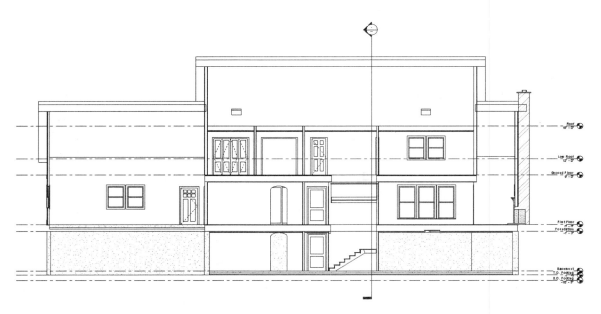

Figure 10-1.7 Longitudinal view – First floor element modified

Exercise 10-2:
Modifying the project model in section view

Again, similar to elevation views, you can modify the project model in section view. This includes adjusting door locations and ceiling heights.

Modifying doors in section view:

In this section you will move a door and delete a door in section view.

1. Open ex10-1.rvt and **Save As ex10-2.rvt**.

2. Open **Cross Section 1** view.

Looking back to Figure 10-1.5, you should notice the wall below the basement stair still extends up through the stair. You will move the wall over in the section view and then note that the adjacent wall and ceilings all automatically adjusted in other views.

3. Use the **Align** tool to make the wall align with the edge of the floor opening. (Figure 10-2.1)

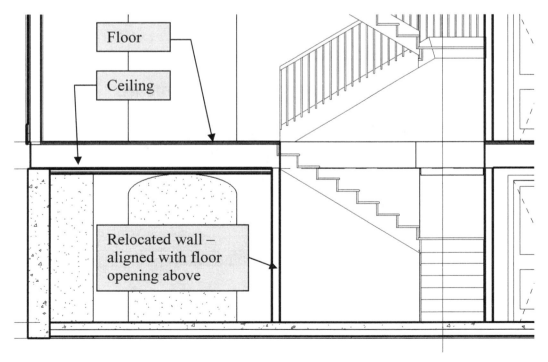

Figure 10-2.1 Cross Section 1 view; wall below stair moved

4. Switch to the **Basement Ceiling Plan** view.

Notice the wall below the stair moved south and the ceiling automatically adjusted as well.

As long as a wall that defines the perimeter of a ceiling does not get deleted or shortened so that the ceiling perimeter is no longer defined, Revit will automatically update the ceiling. If the perimeter does become undefined, you may have to delete the ceiling and add it again.

Next, looking at Figure 10-2.1, notice that the arched opening in the basement is too tall; you will make it shorter. This will require a new type within the Opening-Elliptical Arch family.

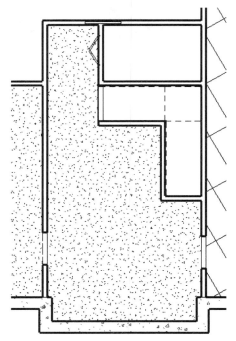

Figure 10-2.2
Basement Floor Plan view; notice changes

5. Switch back to **Cross Section 1**.

6. Select the arched opening in the basement.

7. Click **Properties** → **Edit/New...** → **Duplicate**.

8. Enter **60" x 80"** for the name.

9. Change the **Height** *Parameter* to **6'-8"**. (Figure 10-2.3)

10. Click **OK** twice to close the open dialog boxes.

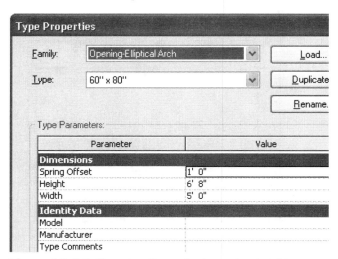

Figure 10-2.3 Changing the arched opening height

Your arched opening should now look like Figure 10-2.4 in Cross Section 1 view. This change would not be apparent in a plan view because the only change was to the opening's height.

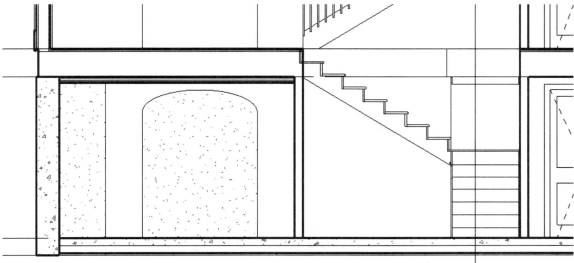

Figure 10-2.4 Cross Section 1 view; arched opening modified

Next you will adjust a ceiling height via your Longitudinal Section view.

11. Switch to **Longitudinal Section** view.

12. At the basement level, select the suspended acoustical ceiling tile system in the family room (lower right room).

Notice that while a ceiling is selected in a section view, a temporary dimension appears. This dimension allows you to directly change the ceiling height.

13. Change the ceiling height to **6'-8"** using the temporary dimension.

Now, as you can see in the section view (Figure 10-2.5), the ceiling has lowered and there is more room for the recessed light fixtures shown.

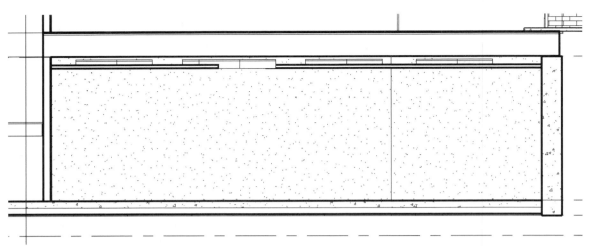

Figure 10-2.5 Cross Section 1 view; ceiling height modified

Adding the strip footing below the foundation wall:

A strip footing, in Revit, is really just a wall type set up as concrete. The width of the wall is set to be the width of your footing, as is the height. Revit provides a predefined wall and plan view (in the residential template); this makes it very easy.

14. Switch to the plan view called **T.O. Footing**
    (**FYI**: T.O. mean "Top Of"; i.e. "top of footing").

15. Select the **Wall** tool and then select **Basic Wall: 24" Footing** from the *Type Selector*.

16. Set the *Loc Line* to **Wall Centerline**.

17. Select **Chain** on the *Options Bar*.

18. Start picking the intersections of wall centerlines, working around the perimeter of the building. Also add a footing below the wall between the garage and the house (you want the 24" footing to be centered on the foundation wall). (Figure 10-2.6)

After drawing the footings, you will notice the interior side of the footing will disappear because it is below the basement floor slab; it is visible in the garage area because it is unexcavated.

> **TIP:** You can double-click on the section bubble to switch to that view. You can also double-click on the target of the level symbol to switch to that plan view.

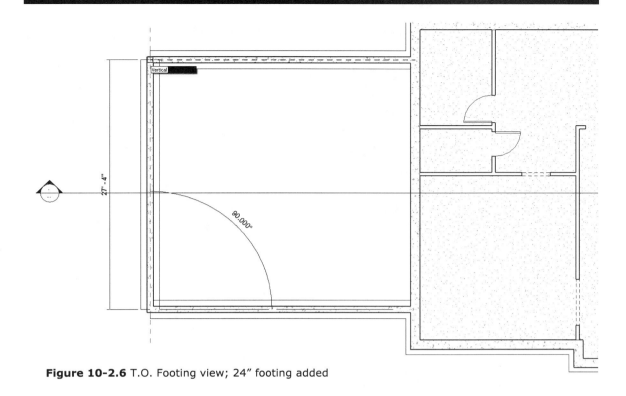

Figure 10-2.6 T.O. Footing view; 24" footing added

19. Switch back to the section view **Cross Section 1**.

As you can see (Figure 10-2.7), the footings have been added.

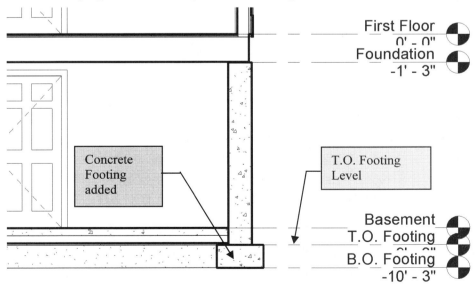

Figure 10-2.7 Cross Section 1 view (modified)

If you look at the properties for the 24" Footing wall type, you can see how the wall was created relative to the T.O. Footing view you drew it in.

Notice the Base and Top Constraints are selected. The Base Constraint is set to **B.O. Footing** and the Top Constraint is set to **Up to level: T.O. Footing**. Also, notice the Unconnected Height is grayed-out, meaning you cannot change this setting, which makes sense seeing as you have defined both the top and bottom of the wall.

The distance between the two levels is 1'-0", so the footing is 1'-0" high.

In Cross Section 1 view, try clicking on the B.O. Footing text (i.e. the 10'-3" text) and change it to -11'-3" and notice the footing size will automatically change. If you try this, make sure you UNDO when you finish this exercise so that your footing is only 1'-0" thick (see image to the right).

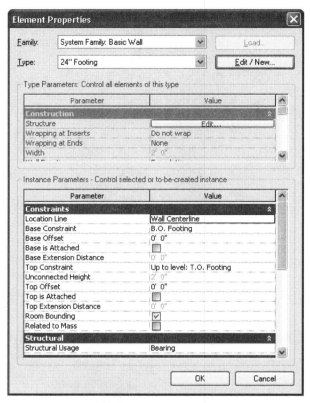

Figure 10-2.8 Properties for 24" Footing wall

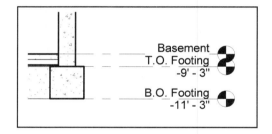

Adding the strip using the continuous footing tool on the Structural Tab:

Revit offers another tool to add continuous strip footings below your foundation wall. On the *Structural* tab (not visible by default; right click *Design Bar* and select Structural) you can select the **Continuous Footing** tool. Then you simply select foundation walls and Revit automatically places a footing, centered, and directly under the wall. If you want to try this option you can delete a portion of footing and use the Structural tool to replace it. If you try this, make sure you *Undo* before proceeding.

Modifying the footing linework in plan view:

Revit has predefined settings for line styles and lineweights. Most of the time, the developers have selected settings that will be satisfactory to most designers, which is great because this saves much time. However, you will occasionally want to tweak these prescribed settings. Revit makes this easy with the **Linework** tool in the *Tools* toolbar. Basically you select the Linework icon, select the desired setting from the Type Selector (e.g. lineweight or line style) and then select the lines to change. This change only applies to the current view.

You will try this next by switching the footing line style to dashed.

20. Switch back to the **T.O. Footing** view.

21. Click the **Linework** tool icon on the *Tools* toolbar.

22. Select **Hidden** from the *Type Selector* (Figure 10-2.9).

Figure 10-2.9 Linework icon selected

23. Click all the visible footing lines; click Ctrl+Z if you pick another line by accident (that will Undo the previous pick).

The footing lines are now hidden (i.e. dashed) in the current view.

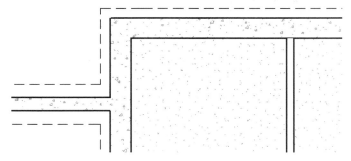

24. Save your project as **ex10-2.rvt**.

Exercise 10-3:
Wall Sections

So far in this lesson you have drawn building sections. Building sections are typically ⅛" or ¼" scale and light on the detail and notes. Wall sections are drawn at a larger scale and have much more detail. You will look at setting up wall sections next.

Setting up the Wall Section view:

1. Open ex10-2 and **Save As ex10-3.rvt**.

2. Switch to the **Cross Section 1** view.

3. From the *View* tab on the *Design Bar*, select the **Callout** tool.

4. Place a **Callout** tag as shown in Figure 10-3.1. *TIP: Pick in the upper left and then in the lower right (don't drag) to place the Callout tag.*

TIP: You can use the Control Grips for the Callout tag to move the reference bubble if desired to move it way from notes/dimensions.

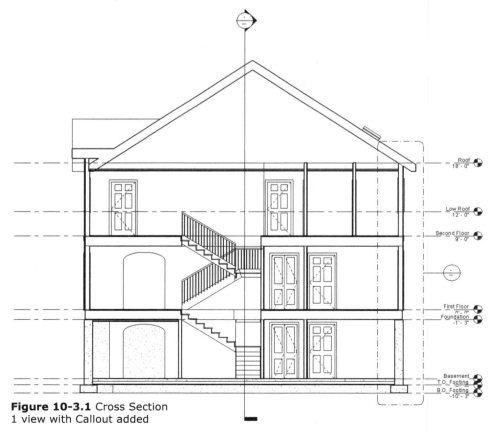

Figure 10-3.1 Cross Section
1 view with Callout added

Notice that a view was added in the *Sections* category of the *Project Browser*. Because *Callouts* are detail references off of a section view, it is a good idea to keep the section view name similar to the name of the callout.

Additionally, *Callouts* differ from section views in that the callout is not referenced in every related view. This example is typical, in that the building sections are referenced from the plans and wall sections are referenced from the building sections. The floor plans can get pretty messy if you try to add too much information to them.

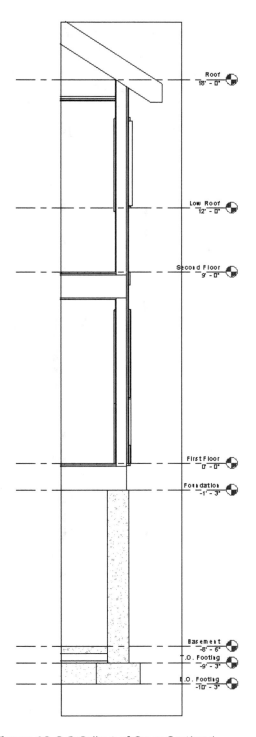

5. Double-click on the reference bubble portion of the *Callout* tag to open the **Callout of Cross Section 1** view. (Figure 10-3.2)

Notice, down on the *View Control* bar, that the scale is set to ½" = 1'-0". This affects the elevation tags (text/targets) and any annotation you add.

6. On the *View Control* bar, set the *View Scale* to **¾" = 1'-0"** and the *Detail Level* to **Fine**.

Notice the Level tags size changed.

If you zoom in on a portion of the *Callout* view, you can see the detail added to the view. The wall's interior lines (i.e. veneer lines) are added and the materials in section are hatched. (Figure 10-3.4 – This is at the second floor line.)

Figure 10-3.2 Callout of Cross Section 1

You can use the Lines tool to add more detailed information to the drawing. For example, you could show the truss, wall base, and flashing. Walls can be configured (and are) to have wall bases, which are set up as sweeps.

As before, you can turn off and adjust the *Crop Region*. If you set the *Detail Level* to Course, you get just an outline of your structure. The *Course* setting is more appropriate for building sections than wall sections. Figure 10-3.3 is an example of the *Course* setting.

    7. **Save** your project as **ex10-3.rvt**.

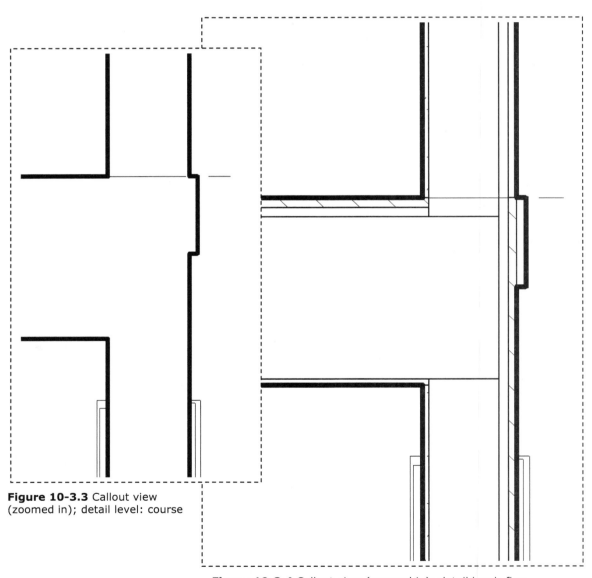

Figure 10-3.3 Callout view (zoomed in); detail level: course

Figure 10-3.4 Callout view (zoomed in); detail level: fine

Exercise 10-4:
Annotation & Detail components

This exercise will explore adding notes, dimensions and detail components to your wall section.

Add notes and dimensions to Callout of Cross Section 1:

1. Open ex10-3.rvt and **Save As ex10-4.rvt**.

2. Switch to **Callout of Cross Section 1** view.

3. Adjust the *View Properties* so the **Crop Region** is not visible.

4. Add the three dimensions as show in **Figure 10-4.1**.
   TIP: Revit lets you dimension to the window opening even though you are not cutting directly through a window in the current view.

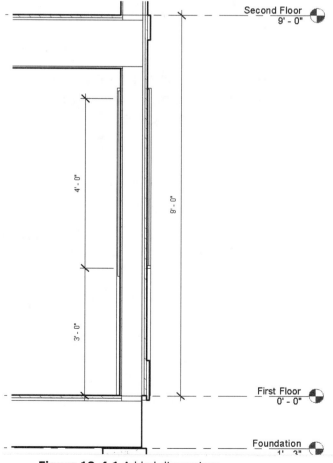

These dimensions are primarily for the carpenters framing the stud wall. Typically, when a window opening is dimensioned in wood stud framing, the dimension has the suffix R.O. This stands for Rough Opening, clearly representing that the dimension identifies an opening in the wall. You will add the suffix next.

5. Select the dimension at the window opening and pick the **Properties** button from the *Options Bar*.

6. Type **R.O.** in the *Suffix* field. (Figure 10-4.2)

7. Click **OK**.

Figure 10-4.1 Added dimensions

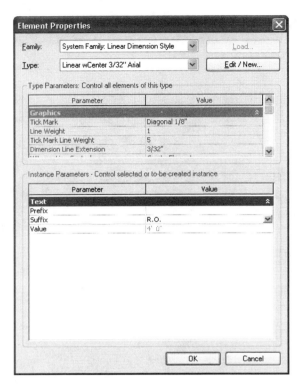

Figure 10-4.2 Selected dimension properties

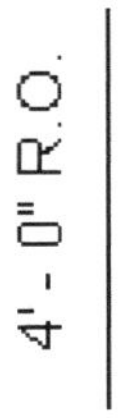

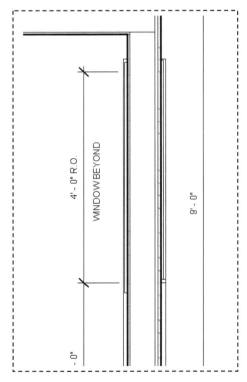

Figure 10-4.3
Dimension with suffix

Figure 10-4.3 shows the dimension with the added suffix.

8. Add the text "WINDOW BEYOND", use Rotate and Move to position it as shown in Figure 10-4.4.

Figure 10-4.4 Text added below dimension

9. Add the notes with leaders shown in Figure 10-4.5.
    (See step 10.)

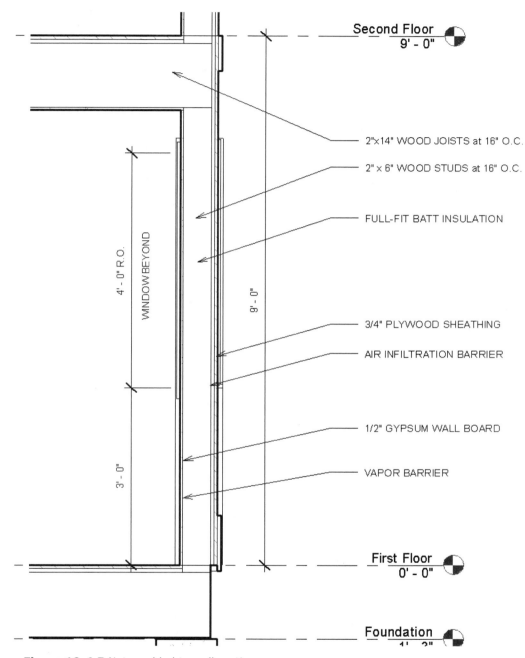

Figure 10-4.5 Notes added to wall section

10. The text style should be set to **3/32" Arial**; it may still be set
    to the last text style you used (¼" Outline Text in this book).

11. Select the text and use the grips and the justification buttons to make the text look like **Figure 10-19**.

Adding detail components:

Revit provides a way in which you can quickly add common detail elements like dimensional lumber in section, anchor bolts, and wall base profiles; these are just a few of the elements loaded with your template file (Figure 10-4.6); many others can be created or downloaded from the internet.

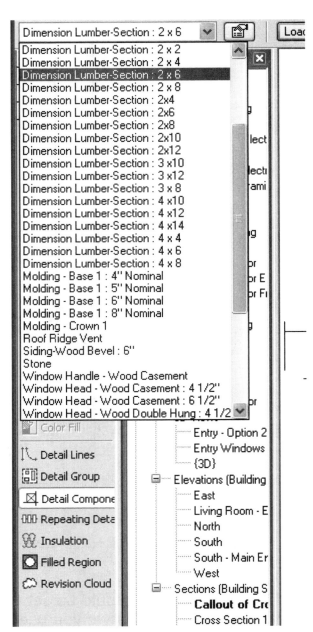

Next you will add just a few *Detail Components* so you have a basic understanding of how this feature is used.

You will add a 2X6 sill plate and double top plate. You will also add the batt insulation symbol in the stud cavity.

12. From the *Drafting* tab, select **Detail Components**.

13. From the *Type Selector*, pick **Dimension Lumber-Section: 2 x 6**.

14. Add the sill plate as shown in Figure 10-4.7.

TIP: When placing the 2x6 you will need to use Rotate and Move to properly position the element. The sill plate should be directly on the ¾" wood sheathing. Note, that the horizontal line that extends through the sill plate is the Level Tag. While placing the component, you can select "rotate after placement" on the Options Bar; this will save you a little time clicking.

Figure 10-4.6 Detail Components; Type Selector open

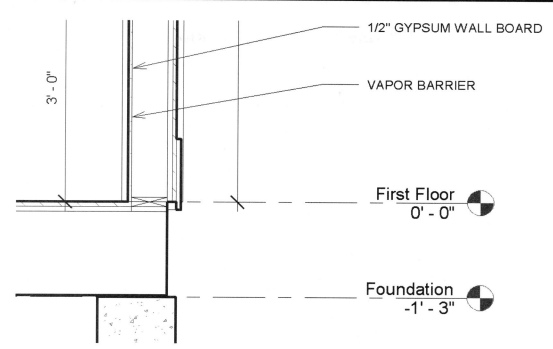

Figure 10-4.7 Callout of Cross Section 1 view; 2 x 6 sill plate added

Detail Components can be copied once you have placed one in the view. This saves the step of rotating the stud in this example.

15. Use **Copy** and **Snaps** to add the double top plate. (Figure10-4.8)

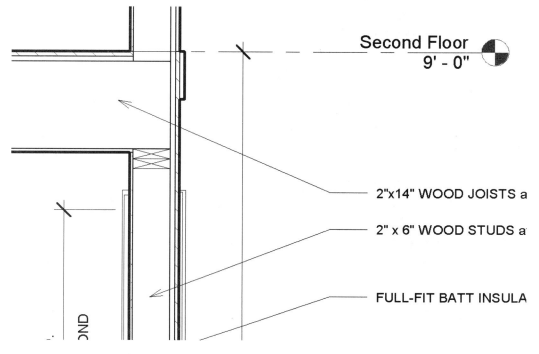

Figure 10-4.8 Callout of Cross Section 1 view; 2 x 6 top plate added

Next you will add the batt insulation.

16. Pick **Insulation** from the *Drafting* tab.

17. Pick the mid-point of the sill plate and the top plate to draw a line that represents the center of the batt insulation symbol (you can pick in either direction). (Figure 10-4.9)

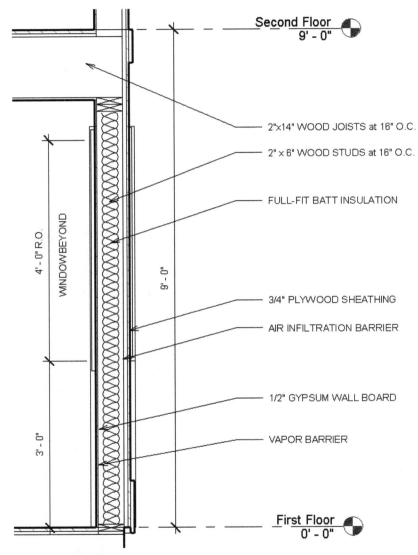

Second Floor
9' - 0"

2"x14" WOOD JOISTS at 16" O.C.

2" x 6" WOOD STUDS at 16" O.C.

FULL-FIT BATT INSULATION

3/4" PLYWOOD SHEATHING

AIR INFILTRATION BARRIER

1/2" GYPSUM WALL BOARD

VAPOR BARRIER

First Floor
0' - 0"

4' - 0" R.O.

WINDOW BEYOND

9' - 0"

3' - 0"

Figure 10-4.9 Callout of Cross Section 1 view; Insulation added

Next you will change the width of the insulation so it fits the space of the stud cavity (i.e. 5½" wide).

18. Click to select the insulation just drawn.

19. Click the Properties button on the Options Bar.

20. Change the **Insulation Width** parameter to **5½″**. (Figure 10-4.10)

21. Click **OK**.

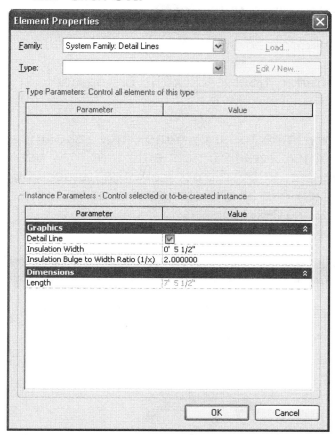

Figure 10-4.10 Insulation properties

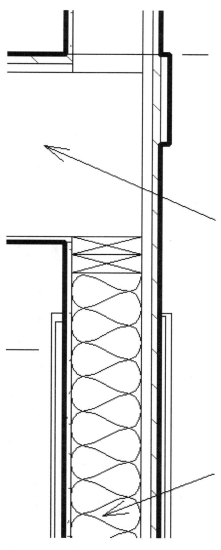

Figure 10-4.11
Insulation width changed

Drawing some elements like this, rather than modeling them 3-dimensionally, can save time and system resources. A file could get very large you tried to model everything. Of course, every time you skip drawing it 3D you increase the chance of error. It takes a little experience to know when to model and when not to model.

FYI: Architectural text is typically all uppercase.

Loading additional detail components:

In addition to the detail components that are preloaded with the template file you started with, you can load more from the Revit content folder on your hard drive.

25. Select **Detail Component** from the *Drafting* tab.

26. Click the **Load** button from the *Options Bar*.

27. Double-click on the *Detail Components* folder.

28. In the *Architectural* sub-folder, double-click the file named **Molding - Crown 3.rfa**.

You now have access to the newly imported crown molding. Take a minute to observe some of the other Detail Components that may be loaded into the project. For example, in the Structural\Steel folder you can load a steel beam Family which has all the standard shapes contained within it as types (Figure 10-4.12).

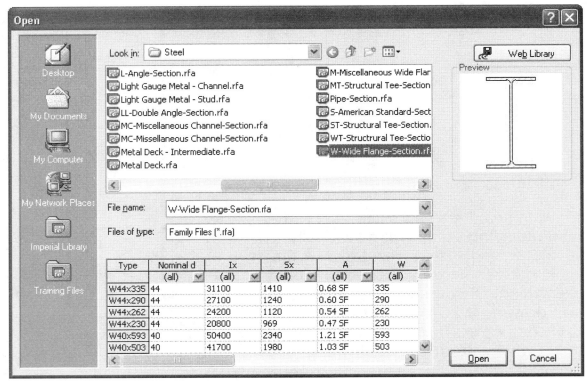

Figure 10-4.12
Load Detail Component dialog (wide flange beams selected)

Using Repeating Details:

Many details and sections have components that occur multiple times, one right after another. Some examples are brick, siding, floor joists in section, etc. Revit provides a tool, called *Repeating Details*, which can basically copy any *Detail Component* the required number of times to fill the space between two pick points you select with the mouse. The spacing of the detail component being copied has to be predefined when setting up the *Repeating Detail* style.

29. Click **Repeating Detail** from the *Drafting* tab.

30. From the *Type Selector*, select **Repeating Detail: Lap Siding 6"**.

For now, you will draw the siding off to one side just to test this feature. Make sure you are within the crop window area or you will not see the results.

31. Pick two points similar to those shown in Figure 10-4.13.

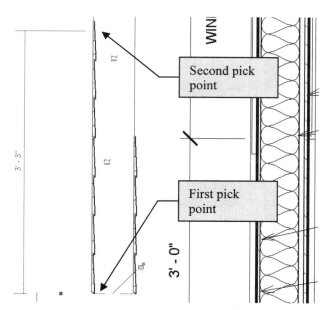

The example to the left shows two Repeating Details added. You can adjust them via the grips or the temporary dimensions.

Figure 10-4.13
Repeating Detail added: lap siding

32. Try adjusting the length of the siding by selecting it and clicking on the dimension text.

One way to quickly study several *Repeating Details* is to download the sample file for repeating details from the Revit content site (Figure 10-4.14).

Figure 10-4.14 Revit's content website

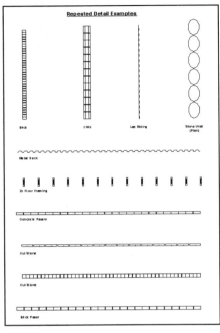

Figure 10-4.15
Sample file downloaded from Revit content website

Once you download the file you can open it and examine how the repeating details work.

Figure 10-4.16
Type Properties for selected repeating detail

The sample file shows Brick, CMU, Lap Siding, Stone Wall (plan), Metal Deck, 2x Floor Framing, Concrete Pavers, Cut Stone, and Brick Paver. (Figure 10-4.15)

Selecting a Repeating Detail and opening its Type Properties dialog reveals the handful of settings associated with it. (Figure 10-4.16)

33. **Delete** the lap siding repeating detail.

34. **Save** your project as **10-4.rvt**.

Self-Exam:
The following questions can be used as a way to check your knowledge of this lesson. The answers can be found at the bottom of this page.

1. The controls for the section mark (when selected) are similar to the controls for the elevation mark. (T/F)

2. Detail Components added in the lesson were 3D. (T/F)

3. The footing drawn in this lesson was a Revit wall type. (T/F)

4. The "Crop Region" is represented by a red line in the section view. (T/F)

5. Use the _____ tool to reference a larger section off a building section.

Review Questions:
The following questions may be assigned by your instructor as a way to assess your knowledge of this section. Your instructor has the answers to the review questions.

1. The visibility of the crop window can be controlled. (T/F)

2. It's not possible to draw a leader (line w/ arrow) with out placing text. (T/F)

3. When a section mark is added to a view, all the other related views automatically get a section mark added to it. (T/F)

4. It is possible to modify objects (like doors, windows & ceilings) in section views. (T/F)

5. You cannot adjust the "depth of view" (width) using the crop window. (T/F)

6. What is the first thing you should do after placing a section tag?

   _____

7. If the text appears to be excessively large in a section view, the view's

   _____ _____ is probably set incorrectly.

8. The abbreviation R.O. stands for _____ _____.

9. Describe what happens when you double-click on the section bubble:

   _____ .

10. Revit provides _____ different leader options within the text command.

Notes:

Lesson 11
Residence: FLOOR PLAN FEATURES::

This lesson explores the various "features," if you will, of a floor plan, such as toilet room layouts (i.e., fixtures and partitions), and cabinets. Additionally, you will look at placing pre-drawn furniture into your project.

Exercise 11-1:
Bathroom layouts

Bathroom layouts involve placing toilets, tubs and sinks. You will start with the 1/2 bath on the first floor.

1. Open ex10-4.rvt and **Save As ex11-1.rvt**.

The template you started with has everything you need to layout this room. You will create the layout shown in Figure 11-1.1.

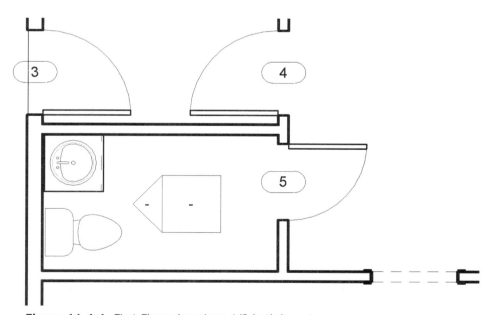

Figure 11-1.1 First Floor plan view: 1/2 bath layout

2. Switch to **First Floor plan view** and zoom in on the 1/2 bath next to the garage and north of the dining room.

The first step will be to place the toilet.

3. With the Component tool selected, pick **Toilet-Domestic-3D** from the *Type Selector*.

4. **Place** the toilet in the middle of the room and then Move and Rotate into place using your snaps. (Figure 11-1.2)

The toilet is 1" out from the wall, which is typical for floor mounted – tank type toilets.

The overlapping letters, C and L, is the symbol for "centerline" which means that the middle of the toilet is 1'-4" from the wall. Actually, if you were to add this dimension, Revit would automatically add this symbol!

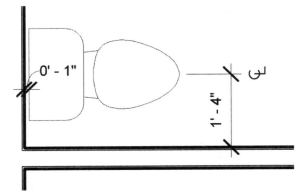

Figure 11-1.2 Toilet location

If you want to make sure the 1'-4" dimension never changes you can add the dimension shown and then click the "Padlock" symbol to *Lock* that dimension. Then, anytime the wall moves, the toilet will move with it. If you want the constraint but do not want the dimension, you can delete the dimension and Revit will ask you if you want to maintain the constraint after the dimension is deleted.

Next, you will add the sink. This consists of three separate components; a base cabinet, a countertop and a sink (which is similar to what you would have in the "real-world"). First you will add the base cabinet.

5. Place a **Vanity Cabinet-Double Door Sink Unit: 24"** using the *Component* tool. (Figure 11-1.3)
    TIP: It is hard to tell which side is the front of the cabinet in plan view, the side the cursor is attached to is the back of the cabinet. If you add it incorrectly you will see the problem when you setup the interior elevation.

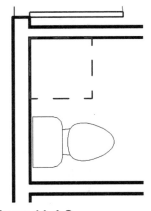

Figure 11-1.3
Vanity base cabinet added

Next you will add the countertop. Revit has one that even has a hole in it for the sink. You will have to adjust the length to fit your cabinet.

6. Using the *Component* tool, place the **Vanity Counter Top w Round Sink Hole: 24"** depth in the middle of the room.

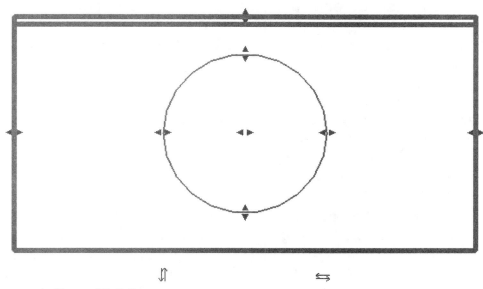

Figure 11-1.4
Vanity counter top temporarily placed in middle of room

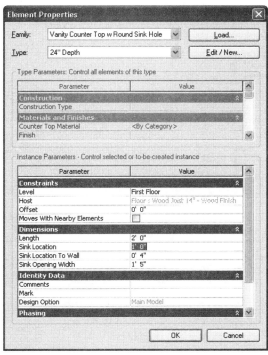

Figure 11-1.5
Vanity counter top properties

You will notice that the counter top is 48" long; you will change this to 24" in a moment. When selected, you see several arrow-grips which allow you to graphically adjust the size of the selected element.

7. With the counter top selected, click **Properties**.

8. Change the *Length* to **2'-0"** and the *Sink Location* to **1'-0"**. (Figure 11-1.5)

9. Click **OK**.

The counter top is now the correct size for your base cabinet.

10. Use **Move** and **Rotate** to position the counter top over the vanity base cabinet; use **Snaps** for accuracy.

11. Again, using the *Component* tool, add **Sink Vanity-Round: 19" x 19"** – use *Snaps* and pick the center point of the hole in the counter top and then *Rotate* the sink 90 degrees.

Your plan should now look like Figure 11-1.1 less the elevation tag, which you will add next.

Interior Elevation view:

Next, you will set up an interior elevation view for the 1/2 Bath. You will also add a mirror above the sinks in elevation view.

12. Place an **Elevation** bubble looking towards the west wall (wall with fixtures on it). (Figure 11-1.6)

13. Rename the new view to: **1/2 Bath** in the *Project Browser*.

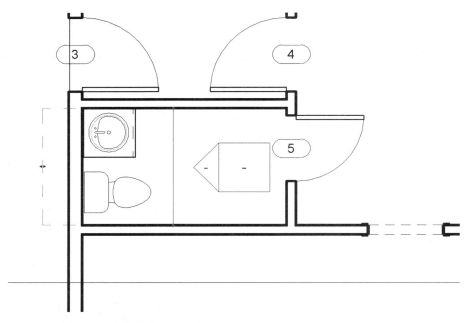

Figure 11-1.6 Elevation tag

14. Switch to the new view; click the Crop Region and drag the bottom up so the floor system does not show. Your view should look like Figure 11-1-7.

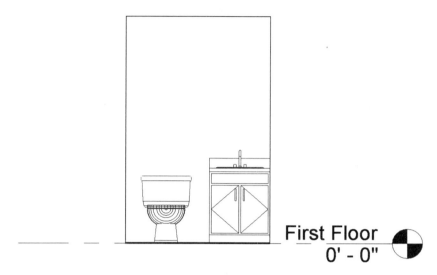

First Floor
0' - 0"

Figure 11-1.7 1/2 Bath elevation

15. Change the *View Scale* to **½" = 1'-0"**.

16. **Load** component *Specialty Equipment\Toilet Room Specialties***Mirror.rfa** from the Online Revit library.

17. Create a *Duplicate* of one of the mirror types just loaded and name it: **Mirror: 24" x 36"**; Change the *Width* to **24"** and the *Height* to **36"**.

18. While in the **1/2 Bath** elevation view, place the mirror as shown in Figure 11-1.8.

19. Add the notes and dimensions shown in Figure 11-1.8.
    TIP: As you are drawing the leaders, Revit will help you align your pick points vertically so the leaders and notes align; just watch for the dashed vertical green line before picking your points.

20. Using the same techniques just covered, layout the two second floor bathrooms per **Figure 11-1.9**; use the tub loaded with the template.

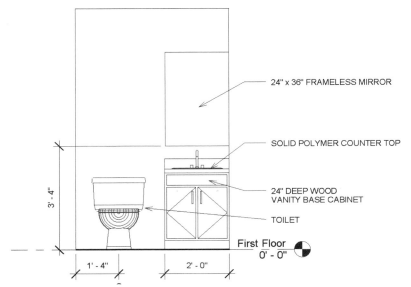

Figure 11-1.8 Interior elevation with mirror and notes added

FYI: Keep in mind that many of the symbols that come with Revit (or any program for that matter) are not necessarily drawn or reviewed by an architect. The point is that the default values, such as mounting heights, may not meet ADA, national, state or local codes. Items like accessible mirrors have a maximum height off the floor to the reflective surface that Revit's standard components may not comply with. However, as you apply local codes to these families, you can reuse them in the future.

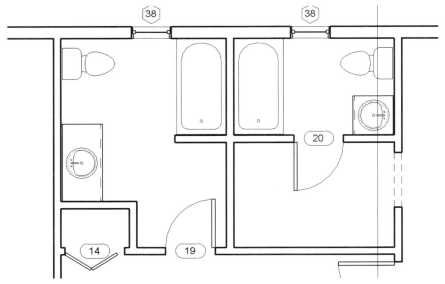

Figure 11-1.9 Second Floor Bath Rooms

21. **Save** your project as **ex11-1.rvt**.

Exercise 11-2:
Kitchen Layout

In this exercise you will look at adding cabinets in the kitchen to your project. As usual, Revit provides several pre-defined families to be placed into the project.

Placing cabinets:

You will add base cabinets and appliances first.

1. Open ex11-1.rvt and **Save As ex11-2.rvt**.

2. Switch to the **First Floor** plan view and zoom into the kitchen.

3. Using the *Component* tool selected, place the Elements as shown in Figure 11-2.1; use *Snaps* for accuracy.
   FYI: *Most of the specified families are already loaded into your project from your template you started with*.

> **FYI:**
> As with other components (i.e. doors and windows), Revit loads several types to represent the most valuable / useful sizes available.
>
> Cabinets typically come in 3" increments with different types (i.e., single or double door unit) having maximum and minimum sizes.

4. Select one of the corner cabinets and change its height to 2'-10½" via *Properties*.
   FYI: *This will match the adjacent cabinets height*.

Next you will add the counter tops to the base cabinets. You will place 5 separate counter top elements (Figure 11-2.2). You will adjust the lengths of the counter tops to fit the cabinets below. The corner units that come with Revit don't quite fit the corner base cabinets; this adjustment is beyond the scope of this book, so it will be left as is.

Once the counter tops are placed, you will use the *Linework* tool to make the line invisible between the corner unit and the sink unit.

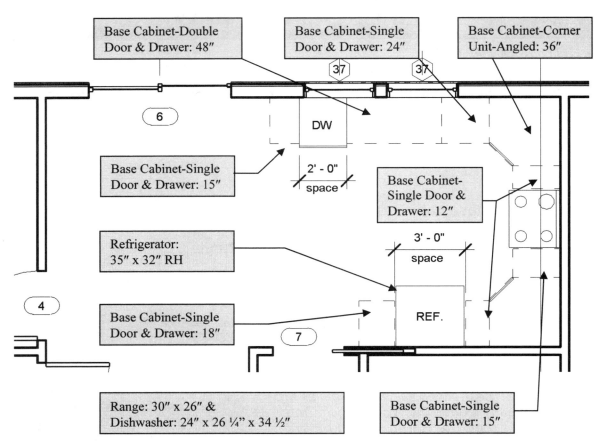

Figure 11-2.1 First Floor Plan View: base cabinets added to kitchen

5. Per the steps listed below, add the counter tops shown in Figure 11-2.2, using *Snaps* to accurately place them.

   a. Load the elements not loaded, in your project, from the *Domestic Kitchen* folder.

   b. Place one of the corner units (use Move, Mirror and Snaps) and then, with the unit selected, drag one of the end grips back to align with the edge of the cabinet below.

   c. Repeat these steps for the other corner.

   d. Place the counter top with the sink hole in the middle of the room; change the length to 48" via Properties.

   e. Add the two straight counter tops as shown.

   f. Verify the *Height* is **3'-0"** via *Properties*.

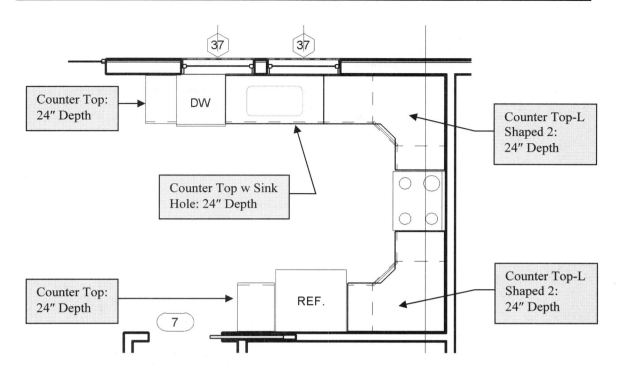

Figure 11-2.2 First Floor Plan View: counter tops added

Next you will use the Linework tool to make the "seam" invisible between the two adjoining counter tops.

6. Select the **Linework** tool's icon from the toolbars. (Figure 11-2.3)

7. Select **<Invisible Lines>** from the *Type Selector*. (Figure 11-2.3)

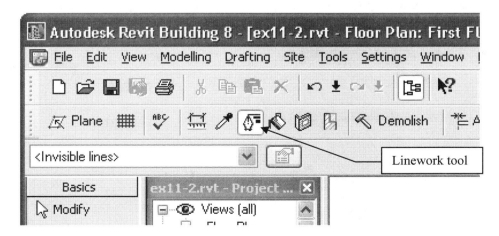

Figure 11-2.3 Linework tool selected; type selector set to <Invisible Lines>

8. Click twice on the "seam" between the two counter tops to make the lines invisible. (Figure 11-2.4)

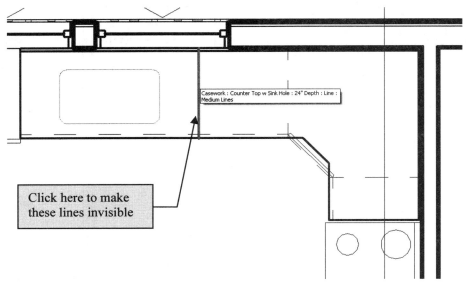

Casework : Counter Top w Sink Hole : 24" Depth : Line :
Medium Lines

Click here to make
these lines invisible

Figure 11-2.4 Linework hovering over lines to become invisible

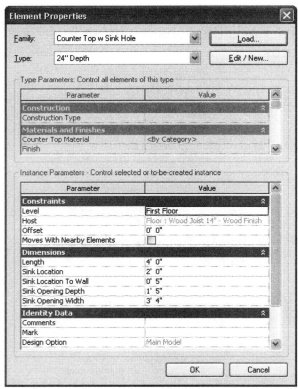

Figure 11-2.5 Properties for Counter top
with Sink Hole: 24" Depth

The counter top now appears monolithic (see Figure 11-2.6). Next you will add the sink and resize the hole in the counter top.

9. Select the counter top with the sink hole and click **Properties**.

10. Change the following: (Figure 11-2.5)
    a. Sink Location to Wall: **5"**
    b. Sink Opening Depth: **1'-5"**
    c. Sink Opening Width: **3'-4"**

TIP: Some settings are per element (i.e. the settings in dialog to the left) and others are per type (i.e. clicking the Edit/New... button); the element settings only affect the selected item.

11. Using the *Component* tool, add **Kitchen Sink-Double: 42" x 21"** to your project – centered on the sink hole. (Figure 11-2.6)

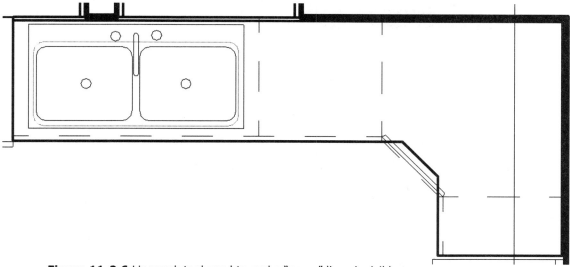

Figure 11-2.6 Linework tool used to make "seam" lines invisible

Placing wall cabinets:

Next you will add a couple wall cabinets to see how they work. They are similar to adding base cabinets, with the exception that they are dependent on walls. You can only place an upper cabinet on a wall, not in the middle of a room like you did with the base cabinets. The base cabinets are setup to sit on the floor and the upper cabinets are assigned a height above the floor (on the wall) via properties.

12. Using the **Component** tool, add the Upper Cabinets as shown in **Figure 11-2.7**:

   a. Move the cursor close to a wall until you see the cabinet (which means Revit found a wall).

   b. Center the upper cabinet above the range, leaving an equal space on each side of the cabinet; these spaces would get a filler that matches the cabinets.

   c. The X's in the cabinets are for clarity of this step only, you do not need to add the lines.

   d. If you have trouble adding the uppers at the pocket door location, add them to the right and move them into place.

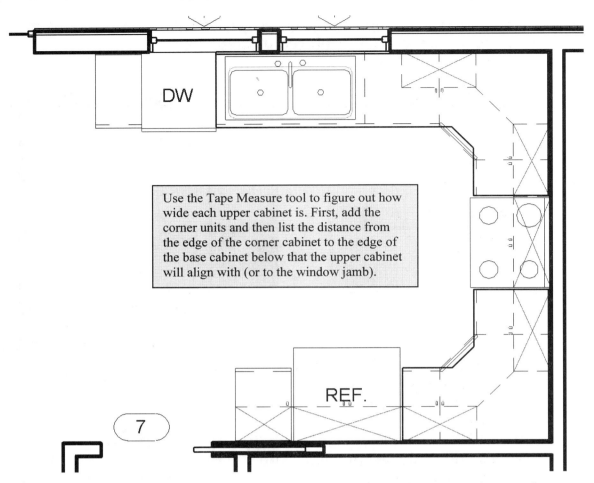

Figure 11-2.7 First Floor plan view: upper cabinets added (X'S added for clarity in this image)

The heights of the cabinets will be adjusted when you get to the interior elevation view; as usual this is gone through Properties.

FYI: Wall cabinets are typically dashed in plan because they are above the floor plan cut plane.

Adding the interior elevation tag:

13. Add an **Elevation Tag** as shown in Figure 11-2.8.

14. Rename the new elevation view to **Kitchen (east)**.

15. Select the <u>square</u> portion of the symbol; the symbol should look like Figure 11-2.9 when selected correctly.

16. Click the north (top) and south (bottom) check-boxes as shown in Figure 11-2.9.

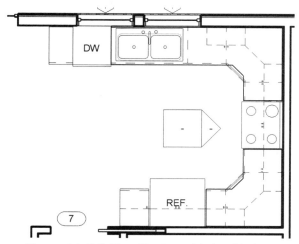

17. **Rename** the two new views to **Kitchen (north)** and **Kitchen (south)**.

The single elevation tag in plan view is associated with the three kitchen elevation views.

Figure 11-2.8 Elevation tag added to kitchen

18. Double-click the triangular portion of the elevation tag pointing towards the east (right) to open that view.

19. Adjust the **Crop Region** so the floor system is not visible.

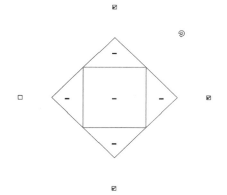

20. Set the *View Scale* to **½" = 1'-0"**; your drawing should look like **Figure 11-10**.

Figure 11-2.9 Elevation tag selected

Depending on where, exactly, you placed the *Elevation Tag*, you may see the side of the refrigerator or a section through it – as in Figure 11-2.10. In this case you will want to adjust the view so you see the cabinets beyond, which is more important to show. You will switch back to the plan view and adjust the views settings.

21. Switch back to **First Floor** plan view.

22. Click to select the east view's triangle (portion of the elevation tag) so you see the *Extent of View / Far Clip Plane* lines.

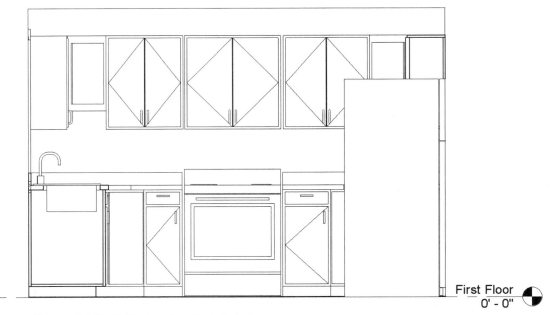

Figure 11-2.10 Kitchen (east); initial view

23. Click and drag to adjust the location of the *Extent of View* line as shown in Figure 11-2.11 (also adjust the Far Clip Plane if necessary.

Notice the Extent of View line has been moved to the east of the refrigerator and the Far Clip Plane is in the middle of the wall.

24. Switch back to the **Kitchen (east)** view and note the changes. (Figure 11-2.12)

You will modify one more thing in this view before moving on. The height of the wall cabinets needs a little work (Figure 11-2.12); the corner cabinets are to short and the cabinet over the range is too tall. You can adjust these in either the plan or elevation view. You will adjust them in the elevation view so you can immediately see the changes.

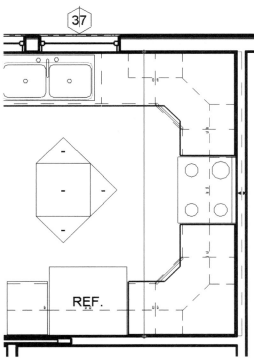

Figure 11-2.11 First Floor; elev tag selected

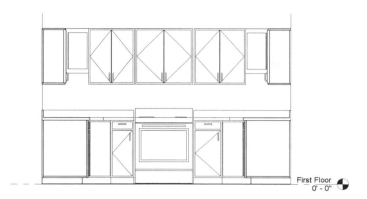

Figure 11-2.12 Kitchen (east) view; view settings adjusted

25. Select both corner upper cabinets and change their height to **2'-6"** and the Elevation to **4'-6"** (2'-6" + 4'-6" = 7'-0" to top). Make these changes via the element and type properties.

26. Select the upper cabinet above the range and change its height to **15"** and the Elevation so the top stays at 7'-0".

27. Search the **Web Library** for a range hood and load/place it.

Your elevation should now look like Figure 11-2.13. Notice that the cabinets on the right and left (which are actually in section) are heavier lines than the cabinet which are in elevation. Again, Revit saves time by assuming the industry standard; however, you can use the Linework tool to change this if you so desire!

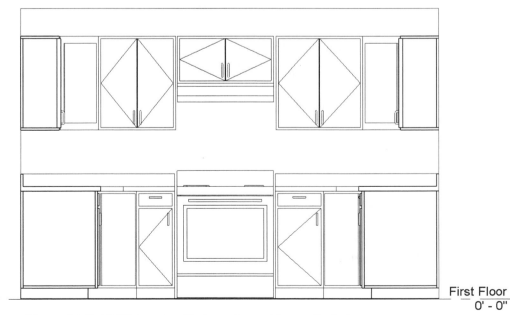

Figure 11-2.13 Kitchen (east) view; view settings adjusted

You will add notes and dimensions to the elevation.

28. Add the notes and dimensions per **Figure 11-2.14**.

29. Select the **Linework** tool and set the *Type Selector* to **Wide lines**. Select the perimeter lines so they stand out more.

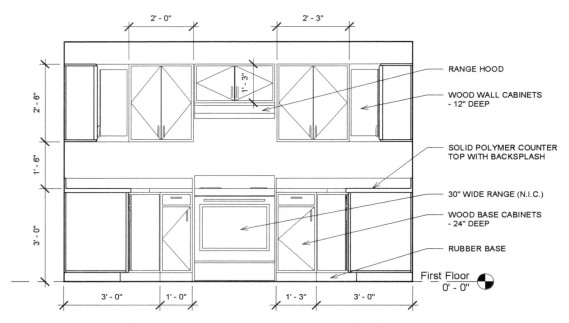

Figure 11-2.14 Interior elevation with annotations

30. Use the **Lines** tool to draw the line behind the refrigerator, indicating the vinyl base.

That's it for the east elevation! Next you will take a quick look at the other two elevations you setup in the kitchen.

31. Switch to the **Kitchen (north)** view.

32. Adjust the *Crop Region* and the perimeter lines per the previous elevation. (Figure 11-2.15)

33. Change the *View Scale* to **½" = 1'-0"**.

You will notice right away that the windows are in conflict with the cabinets and the counter top does not extend over the dishwasher. Both of these required changes can be made in either the plan or the elevation; again, you will make changes in the elevation view so you can see the results.

First you will fix the windows. An error like this could easily be overlooked using a traditional CAD program because the exterior elevation and the interior elevations are separate / independent drawings files; so you could have fixed the height of the windows in the interior elevation, but forget to change the exterior elevation (where the windows are typed and ordered from). In Revit, it is not even possible to have one window be two different sizes in the same project file.

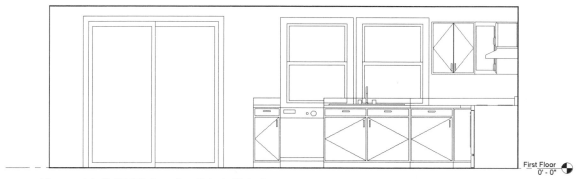

Figure 11-2.15 Kitchen (north); initial view

34. Select the two windows in the **Kitchen (north)** view.

35. Via *Properties*, create a new window *Type* named
   Double Hung with Trim: 36" x 42":
   a. Set the *Height* to **3'-6"** (from 4'-0")
   b. See the *Sill Height* to **3'-6"** (From 3'-0")

Now the windows do not conflict with the cabinets. Now you will adjust the counter top.

36. In the Kitchen (north) view, zoom in on the dishwasher area.

37. Select the **Counter Top** over the **15"** wide base cabinet (to the left of the range).

38. On the right side of the selected *Counter Top*, click and drag (see Step 38) the arrow-grip that points to the right and drag it to the right.

39. Drag the counter towards the east until it aligns with the counter over the sink base.

40. Switch to plan view and use the **Linework** tool to thicken the perimeter lines again (in elevation, you can also make the joint between counter tops (in elevation) disappear.

Your modified Kitchen (north) should look like Figure 11-2.16.

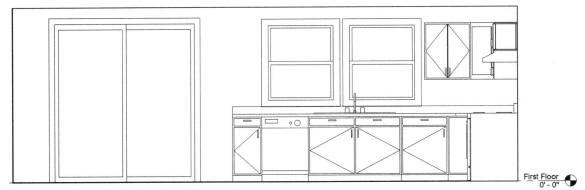

Figure 11-2.16 Kitchen (north); modified view

41. Open **Kitchen (south)**.

42. Modify the **Crop Region** and the perimeter linework.

You cannot see the refrigerator doors, you new realize that the refrigerator was inserted backwards (Figure 11-2.17), you will fit this next.

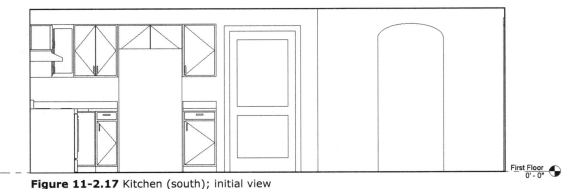

Figure 11-2.17 Kitchen (south); initial view

43. **Rotate** the refrigerator (in plan view) 180 degrees.

44. Modify the upper cabinet above the refrigerator to be 12" tall (with the top at 7'-0"). **TIP:** *If this cabinet where used anywhere else in the project you would have had to make a duplicate before changing the height; otherwise, all the cabinets would have changes as well (because it is a type parameter you are changing, not an element parameter).*

Often, you have a portion of a wall that does not have very much on it, so it is not worth elevating. In the <u>Kitchen (south)</u> view you will "crop" a portion of the wall on the right.

45. Click the **Crop Region** and drag the right side of the *Crop Region* rectangle towards the left so only the portion of elevation shown in Figure 11-2.18 is visible.

Your Kitchen (south) view should now look like Figure 11-2.18.

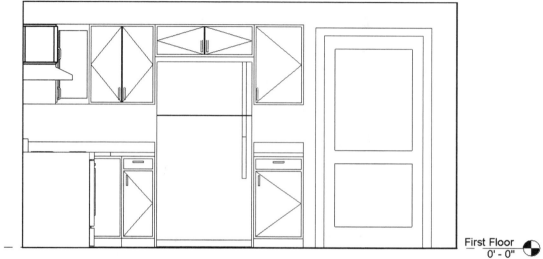

Figure 11-2.18 Kitchen (south); modified view

That is for the kitchen! With a little practice you could design a kitchen in very short order.

46. **Save** your project as **ex11-2.rvt**.

TIP: You may have noticed on the Web Library, a listing for the American Woodwork Institute. This section has Revit families for most of the industry standard cabinets. You can download the cabinets from their website. You can even print out a small flyer that shows an image each cabinet that can be downloaded.

Exercise 11-3:
Furniture

This lesson will cover the steps required to lay out furniture. The processes are identical to those previously covered for toilets and cabinets. First you will start with the office on the second floor.

Loading the necessary families:

1. Open ex11-2.rvt and **Save As ex11-3.rvt**.

2. Select the *Component* tool and load the following items into the current project:

    Local Files *(i.e., on your hard drive)*
    a. **Work Station Desktops** (Furniture System)
    b. **Sofa-Pensi** (Furniture)
    c. **Chair-Breuer** (Furniture)
    d. **Chair-Executive** (Furniture)
    e. **Table-Round** (Furniture)

    Online Files *(i.e., Revit's Web Library on internet)*
    f. Specialty Equipment\Office Equipment**Copier-Floor**

These files represent various predefined families that will be used to design the second floor office.

> **TIP:** You can set the View mode for the Open dialog box (which is displayed when you click *Load from Library*. One option is Thumbnail mode; this displays a small thumbnail image for each file in the current folder. This makes it easier to see the many symbols and drawings that are available for insertion.
>
>
> **Above:** View set to List mode
>
> **Right:** View set to Thumbnail mode

Designing the office furniture layout:

3. Switch to the **Second Floor** view.

4. Place the furniture as shown in Figure 11-3.1.
   TIP: *Use snaps to assure accuracy; use rotate and mirror as required.*

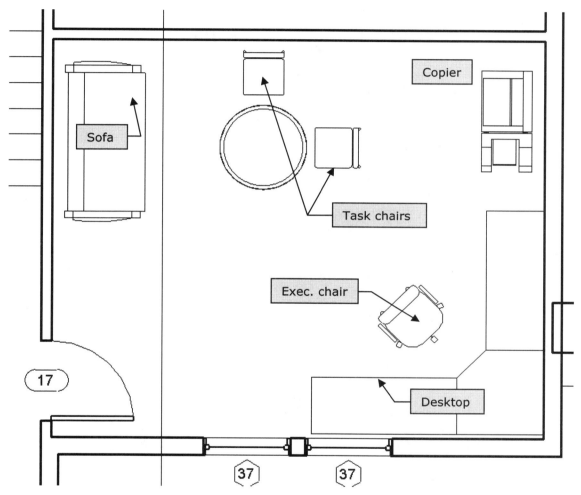

Figure 11-3.1 Second Floor – Office furniture layout

Look at the properties for the desktop to see the Desktop Height setting which controls how high the surface is above the current floor.

You may want to browse through the component library and the web library for miscellaneous residential furniture. You can add dining room tables and chairs, book shelves, washer and dryer, etc. You instructor may require you to complete this suggested step. An example of a few items placed can be seen in Figure 11-3.2.

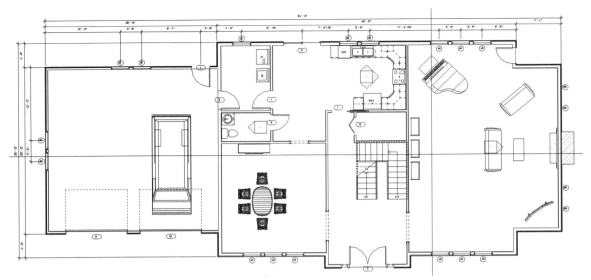

Figure 11-3.2 First Floor – one possible furniture layout

3D view of office layout:

Next you will look at a 3D view of your office area. This involves adjusting the visibility of the roof and skylights.

5. Switch to the **3D** view.

6. Right click anywhere in the drawing window and select **View Properties** from the pop-up list.

7. Click **Edit** next to the *Visibility* parameter.

8. Uncheck the *Ceiling* and the *Roof* category and click **OK** twice.

The ceilings and roof should not be visible now. However, you should still see the light fixtures floating in space. You will make those disappear next.

9. Select one of the 2' x 4' light fixtures floating above the office.

10. Click the **Hide/Isolate** tool from the *View* toolbar.

You should see the menu shown in **Figure 11-3.3** show up next to the Hide/Isolate icon. This allows you to isolate an object (so it's the only thing on the screen) or hide it (so the object is temporarily not visible).

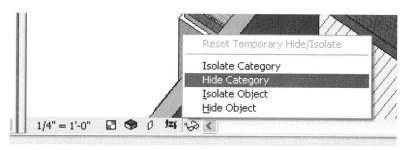

Figure 11-3.3 Hide/Isolate popup menu

11. Click **Hide Category** in the menu. (Figure 11-3.3)
     FYI: This makes all the lights hide; you could leave it at "Hide Object" and select each light.

12. Adjust your 3D view to look similar to **Figure 11-3.4**; try looking at the room from other angles as well.

You will now restore the original visibility settings for the 3D view. Notice the *Hide/Isolate* icon now has a red background; which means something in the project is temporarily hidden from view.

13. Click the Hide/Isolate icon and then select **Reset Temporary Hide/Isolate** from the popup up menu.

14. Reset the 3D view's visibility settings so the ceilings and roof are visible.

15. **Save** your project as **ex11-3.rvt**.

FYI: Not all families found online are drawn in full 3D. Some may have been brought over from a CAD program where all the symbols are drawn in 2D only.

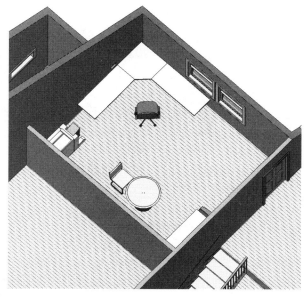

Figure 11-3.4 3D view with ceiling/roof/lights hidden

Online Content:

A few locations on the internet provide additional content for use in Revit. Some is free and some is not. Hopefully product manufacturers will start providing content based on the products they make, making it easier for people to include that manufacturer's product in their project (both the virtual and real projects).

You have already spent a little time looking at Revit's online content library (revit.autodesk.com/library/html/). You should spend some more time there so you know what is generally available. This will help you to reduce duplicated effort.

The Modern Medium 8.0 Library, available in Revit's content library, has a ton of furniture you can use in your project. They have Accurender materials pre-assigned so they are ready to render (see sample screen shot below).

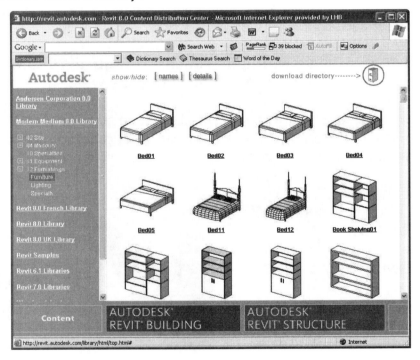

The following sites also contain content that can be downloaded:
- www.revitcity.com
- www.augi.com
- www.revitdrop.com

You should occasionally search the internet to see if additional content becomes available. You can do a Google search for "revit content"; make sure to include the quotation marks. The rendering content, such as that offered by www.archvision.com, will be covered in chapter 13.

Exercise 11-4:
Adding Guardrails

This lesson will cover the steps required to layout guardrails. The steps are similar to drawing walls; you select your style and draw its path.

Adding a guardrail to the second floor stair area:

1. Open ex11-3.rvt and **Save As ex11-4.rvt**.

2. Switch to **Second Floor** plan view.

3. From the *Modeling* tab select **Railing**.

4. **Zoom** into the stair area.

At this point you will draw a line representing the path of the guardrail. The railing is offset to one side of the line, similar to Walls. However, you do not have the *Loc Line* option as you do with the Wall tool, so you have to draw the railing in a certain direction to get the railing to be on the floor and not hovering in space just beyond the floor edge.

5. Draw a line along the edge of the floor as shown in **Figure 11-4.1**.

   TIP: Select Chain from the Options Bar to draw the railing with fewer picks.

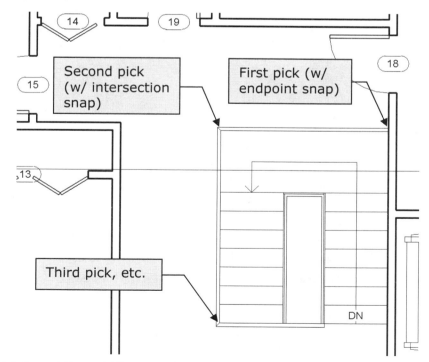

Figure 11-4.1 Second Floor: Adding guardrail

6. Click **Finish Sketch** from the *Design Bar*.

The railing has now been drawn. In the next step you will switch to a 3D view and see how to quickly change the railing style. This will also involve changing the height of the railing. Most building codes require the railing height be 42" when the drop to the adjacent surface is more than 30"; this is called a guardrail.

7. Switch to the **3D** view.

8. Turn off the ceilings and roof as previously reviewed and then zoom into the railing just added to the second floor. (Notice the railing style – Figure 11-4.2.)

Figure 11-4.2 Added railing – 3D view

9. Select the railing. You may have to use the Tab key to cycle through the various selection options.

10. With the railing selected, select the other railing type available in the *Type Selector* on the *Options Bar*. When finished make sure **Railing: Handrail – Rectangular** is selected.

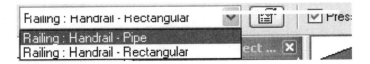

When the alternate style was selected, your railing should have looked like Figure 11-4.3 (even the railing for the stair was changed – by a similar step). This style is too contemporary for this design so you will not use it at this time. FYI: you can change the height via *Properties*.

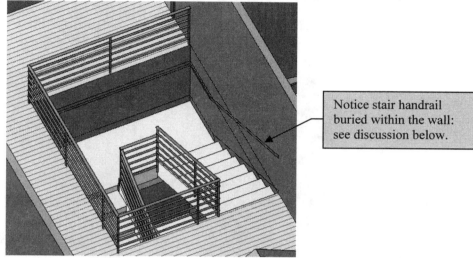

Notice stair handrail buried within the wall: see discussion below.

Figure 11-4.3 Railing with alternate style

Editing the Sketch Lines for the stair railing:

You will notice, in the image above (Figure 11-4.3), that a portion of the railing added with the stair is buried in the wall. You will edit the railing sketch lines to remove this conflict.

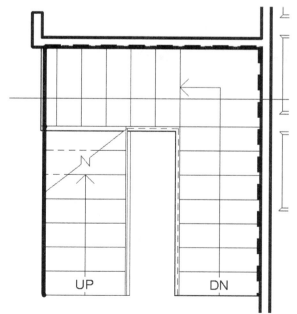

Figure 11-4.4 Edit railing sketch lines

11. Switch to the **First Floor** plan view.

12. Select the outer railing for the upper stair (use the Tab key to cycle through your selection options).

13. Click the **Edit** button on the *Options Bar*.

14. Delete the sketch lines on the north and east walls (shown dashed in Figure 11-4.4)

15. Select **Finish Sketch** on the *Design Bar* when finished.

Switching back to the 3D view you notice the railing lines in the walls are gone (Figure 11-4.5). Revit is consistent in how many 3D components are defined and edited with 2D sketch linework; editing the stringer extensions is a similar process. This consistency help reduce the learning curve.

Figure 11-4.5 Railing conflict resolved

The railing tool has the ability to create fairly complex railings. The image below (Figure 11-4.6) shows the interface for setting up the baluster placement. This kind of advanced control is beyond the scope of this textbook.

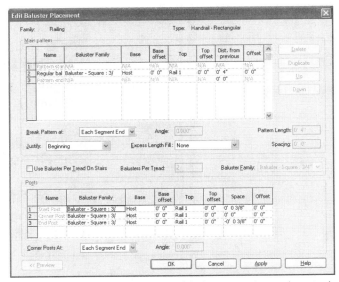

Figure 11-4.6 Dialog to control baluster and newel post placement

Make sure to examine the railing sample file available on Revit's online content library (revit.autodesk.com/library/html/). You can download this file, open it, select a railing and view its properties to see how it works.

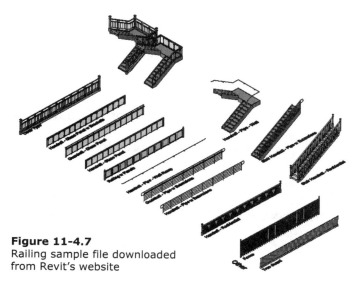

Figure 11-4.7
Railing sample file downloaded
from Revit's website

You can Copy/Paste a railing style from this drawing into one of your project files. Then you select your railing and pick the newly imported one from the *Type Selector*. This process was done to achieve the image below (Figure 11-4.8). Notice the traditional balusters and newel post.

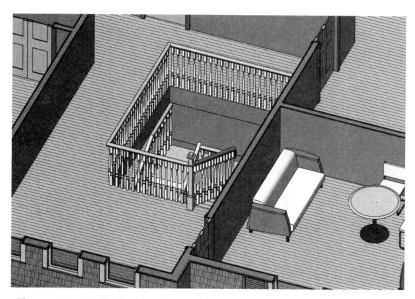

Figure 11-4.8 Optional railing configuration

16. **Save** your project as **ex11-4.rvt**.

Self-Exam:
The following questions can be used as a way to check your knowledge of this lesson. The answers can be found at the bottom of this page.

1. The bath room fixtures are preloaded in the template file. (T/F)

2. You do not need to be connected to the internet when you click on the *Load from Web* button in the Load from Library dialog box. (T/F)

3. The Revit items are not always in compliance with codes. (T/F)

4. You can use Crop Region to crop a portion of wall that is blank. (T/F)

5. Use the _____ tool to copy fixtures to other floors

Review Questions:
The following questions may be assigned by your instructor as a way to assess your knowledge of this section. Your instructor has the answers to the review questions.

1. Revit provides several different styles of vanity cabinets for placement. (T/F)

2. Most of the time Revit automatically updates the ceiling when walls are moved, but occasionally you have to manually make revisions. (T/F)

3. It is not possible to draw dimensions on an interior elevation view. (T/F)

4. Cabinets typically come in 2" increments. (T/F)

5. Base cabinets automatically have a countertop on them. (T/F)

6. What can you adjust so the floor system does not show in an int. elevation?

   _____

7. You can use the _____ _____ to control element visibility

8. What is the current size of your Revit Project? _____

9. What should you use to assure accuracy when placing cabinets?

   _____

10. You use the _____ tool to make various components temporarily invisible.

Lesson 12
Residence: SCHEDULES::

You will continue to study the powerful features available in Revit. The ability to create parametric schedules is very useful. For example, you can delete a door number on a schedule and Revit will delete the corresponding door from the plan.

Exercise 12-1:
Room and Door Tags

This exercise will look at adding room tags and door tags to your plans. As you insert doors, Revit adds tags to them automatically. However, if you copy or mirror a door (or add one in elevation) you can lose the tag and have to add it.

Adding Room Tags to the First Floor:

You will add a Room Tag to each room on your first floor plan.

⊝ Room Tag

1. From the *Drafting Tab* on the *Design Bar*, select **Room Tag**.

Placing a room tag is similar to placing a ceiling in the reflected ceiling plan; as you move your cursor over a room, the room (perimeter) highlights. When the room you want to place a Room Tag in is highlighted, you click to place the tag in that room.

2. Place your cursor within the Entry area and place a *Room Tag*. (Figure 12-1.1)

By default, Revit will simply label the space 'Room' and number it '1.' You will change these to something different.

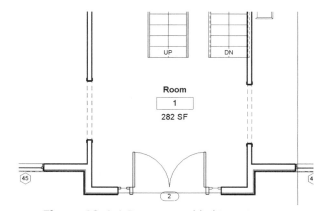

Figure 12-1.1 Room tag added to entry

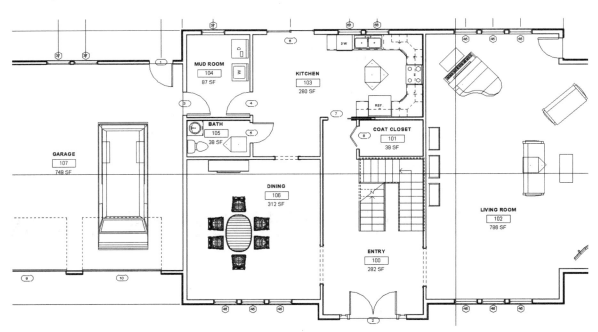

Figure 12-1.2 Room tags – First Floor

3. Press **Esc** or select **Modify** to cancel the *Room Tag* command.

4. Click on the *Room Tag* you just placed to select it.

5. Now click on the room name label to change it; enter **Entry**.

6. Now click on the room number to change it; enter **100**.

7. Enter *Room Tags* for each room on the first floor, incrementing each room number by 1. (Figure 12-1.2)

Adding Room Tags to the Second & Third Floors:

8. Add *Room Tags* to Second Floor. The numbering should start with 200; do **not** change the room names. (Figure 12-1.3)

9. Add *Room Tags* to Basement Floor plan. The numbering should start with 001; **do** change the room names. (Figure 12-1.4)

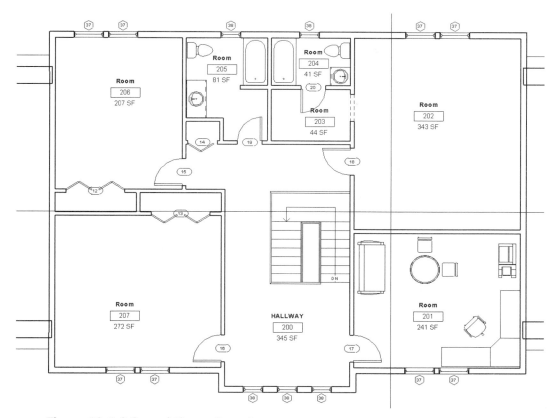

Figure 12-1.3 Second Floor – Room tags

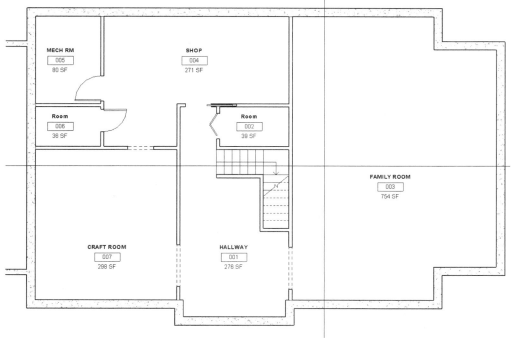

Figure 12-1.4 Basement – Room tags

Adding Door Tags:

Next you will add Door Tags to any doors that are missing them. Additionally, you will adjust the door numbers to correspond to the room numbers.

Revit numbers the doors in the order they are placed into the drawing. This would make it difficult to locate a door by its door number if door number 1 was on the first floor and door number 2 was on the second floor, etc. Typically, a door number is the same as the room the door swings into. For example, if a door swung into a room numbered 104, the door number would also be 104. If the room had two doors into it, the doors would be numbered 104A and 104B.

Figure 12-1.5
Drafting tab

10. Switch to **First Floor** plan view.

11. Click the **Tag** button on the *Drafting* tab. (Figure 12-1.5)

Notice as you move your cursor around the screen Revit displays a tag, for items that can have tags, when the cursor is over it. When you click the mouse is when Revit actually places a tag.

12. **Uncheck** the **Leader** option on the *Options Bar*.

13. Place a door tag for each door that does not have a tag, do this for each level.
    TIP: This should mainly be the doors in the basement and the one added in elevation in the living room.

14. Renumber all the door tags to correspond to the room they open into; do this for each level. (Figure 12-1.6 is an example of 1st floor and 12-1.7 of the 2nd floor.)
    Remember *to click Modify, select the Tag and then click on the number to edit it.*

TIP:
Tag All Not Tagged...
This tool allows you to quickly tag all the objects of a selected type (e.g. doors) at one time.

After selecting the tool, you select the type of object from a list and specify whether or not you want a leader. When you click OK, Revit tags all the untagged doors in that view.

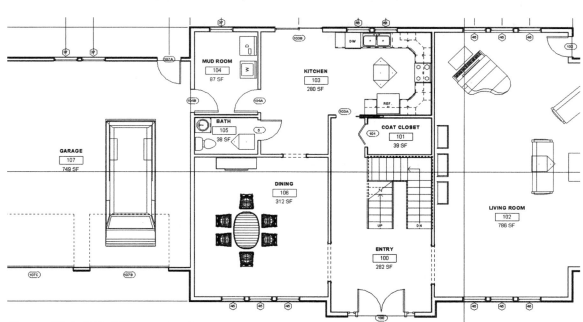

Figure 12-1.6 First Floor – door tags

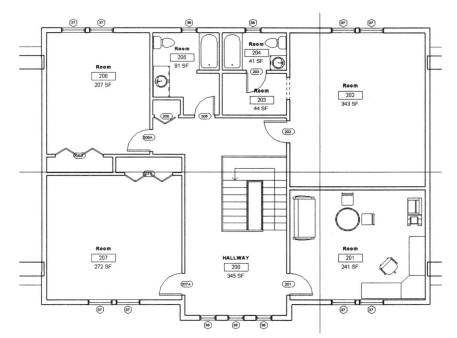

Figure 12-1.7 Second Floor – door tags

15. **Save** your project as **ex12-2.rvt**.

Exercise 12-2:
Generate a Door Schedule

This exercise will look at creating a door schedule based on the information currently available in the building model (i.e., the tags).

Create a Door Schedule view:

A door schedule is simply another view of the building model. However, this view displays numerical data rather than graphical data. Just like a graphical view, if you change the view it changes all the other related views. For example, if you delete a door number from the schedule, the door is deleted from the plans and elevations.

1. Open ex12-1.rvt and **Save As ex12-2.rvt**.

2. Select the **Schedule/Quantities** button from the *View* tab on the *Design Bar*.

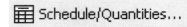

3. Select **Doors** under *Category* and then click **OK**. (Figure 12-2.1)

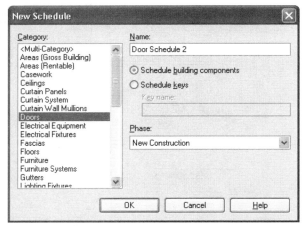

Figure 12-2.1 New Schedule dialog

FYI:
The template you started with already has a door schedule setup and ready to go. So the first door you added was already scheduled. However, you will go thru the steps of setting up a schedule so you understand how a schedule is created.

You should now be in the *Schedule Properties* dialog where you specify what information is displayed in the schedule, how it is sorted and the text format.

4. On the **Fields** tab, *Add* the information you want displayed in the schedule. Select the following: (Figure 12-2.2)
    a. Mark
    b. Width
    c. Height
    d. Frame Material
    e. Frame Type
    f. Fire Rating

As noted in the dialog, the fields added to the list on the right are in the order they will be in the schedule view. Use the *Move Up* and *Move Down* buttons to adjust the order.

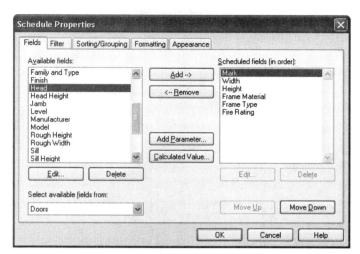

Figure 12-2.2 Schedule Properties - Fields

5. Do **NOT** Click OK *(you need to adjust one more thing)*.

6. On the **Sorting/Grouping** tab, set the schedule to be sorted by the **Mark** (i.e., door number) in *Ascending* order. (Figure 12-2.3)

TIP:

*The **Formatting and Appearance tabs** allow you to adjust how the schedule looks. The formatting is not displayed until the schedule is placed on a plot sheet.*

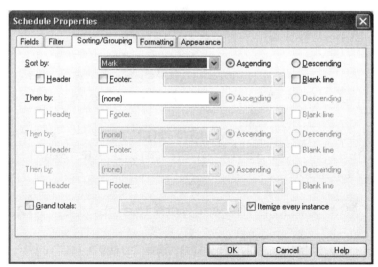

Figure 12-2.3 Schedule Properties – Sorting

7. Click the **OK** button to generate the schedule view.

You should now have a schedule similar to Figure 12-2.4.

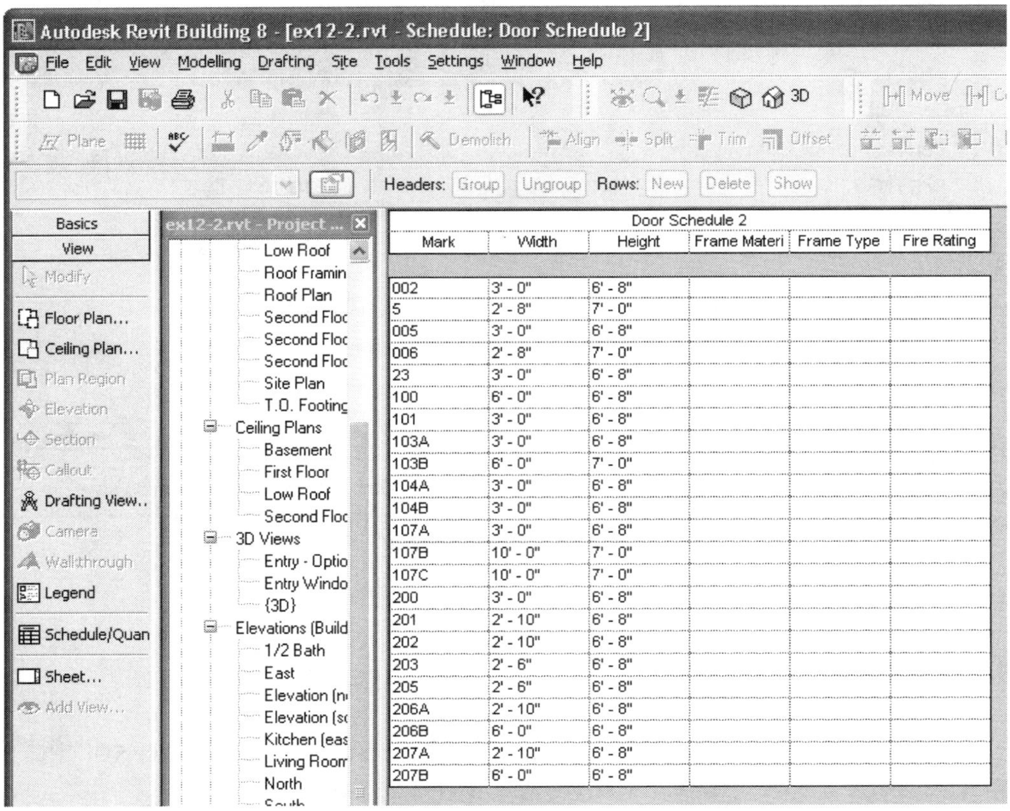

Figure 12-2.4 Door schedule view

Next you will see how deleting a door number from the schedule deletes the door from the plan.

8. Switch to the **Door Schedule 2** view, under Schedules/ Quantities in the *Project Browser*.

Next you will delete door 102 from the door schedule view; this was the door you added in an interior elevation – the exterior door in the living room.

9. Click in the cell with the number **102**.

10. Now click the **Delete** button from the *Options Bar*. (Figure 12-2.5)

Figure 12-2.5 Options Bar for the door schedule view

You will get an alert. Revit is telling you that the actual door will be deleted from the project model. (Figure 12-2.6)

    11. Click **OK** to delete the door. (Figure 12-2.6)

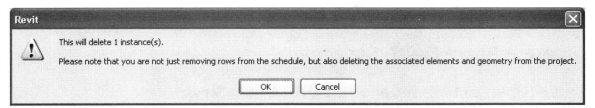

    Figure 12-2.6 Revit alert message

    12. Switch back to the **First Floor** view and notice that door 102 has been deleted from the project model (Figure 12-2.7).

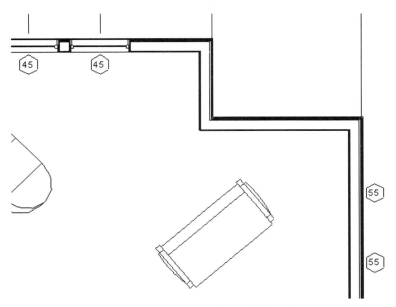

    Figure 12-2.7 First Floor plan view: door 102 has been removed

    13. **Save** your project as **ex12-2.rvt**.

TIP:
You can also change the door number in the schedule and the even the size; however, changing the size actually changes the door family which affects all the doors that size and style.

Exercise 12-3:
Generate a Room Finish Schedule

In this exercise you will create a Room Finish schedule. The process is similar to the previous exercise.

Create a Room Finish Schedule:

1. Open ex12-2.rvt and **Save As ex12-3.rvt**.

2. Select the **Schedule/Quantities** button from the *View* tab on the *Design Bar*.

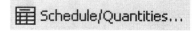

3. Select **Room** under *Category* and then click **OK**. (Figure 12-3.1)

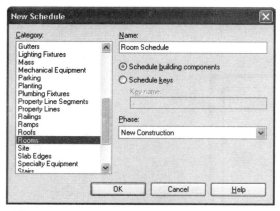

Figure 12-3.1 New Schedule dialog

4. In the **Fields** tab of the *Schedule Properties* dialog, add the following fields to be scheduled (Figure 12-3.2):
    a. Number
    b. Name
    c. Base Finish
    d. Floor Finish
    e. Wall Finish
    f. Ceiling Finish
    g. Area

Area is not typically listed on a room finish schedule. However, you will add it to your schedule to see the various options Revit allows.

5. On the **Sorting/Grouping** tab set the schedule to be sorted by the **Number** field.

6. On the **Appearance** tab, select **Bold** for the header font. (Figure 12-3.3)

7. Select **OK** to generate the **Room Schedule** view.

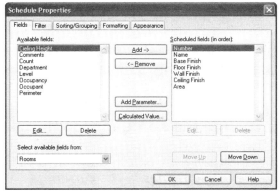

Figure 12-3.2
Schedule Properties - Fields

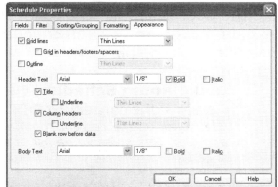

Figure 12-3.3
Schedule Properties - Appearance

Place cursor here to resize the column

Figure 12-3.4 Room Schedule view

Your schedule should look similar to the one to the left (Figure 12-3.4).

8. Resize the **Name** column so all the room names are visible. Place the cursor between the *Name* and *Base Finish* and drag to the right until all the names are visible.

The formatting (i.e., Bold header text) will not show up until the schedule is placed on a plot sheet.

Modifying & Populating a Room Schedule:

Like the door schedule, the room schedule is a tabular view of the building model. So you can change the room name or number on the schedule or in the plans.

9. In the **Room Schedule** view, change the name for room **201** (lower right corner) to **OFFICE**.
TIP: Click on the current room name and then click on the down-arrow that appears. This gives you a list of all the existing names in the current schedule; otherwise you can type a new name.

10. Switch to the **Second Floor** plan view to see the updated room tag.

You can quickly enter finish information to several rooms at one time. You will do this next.

11. In the Second Floor plan view, select the Room Tags for all bedrooms (3 total).
REMEMBER: Hold the Ctrl key down to select multiple objects.

12. Click the **Properties** button on the *Options Bar*.

The Parameters listed here are the same as the *Fields* available for display in the room schedule. When more than one tag is displayed and a parameter is not the same (e.g., different names), that value field is left blank. Otherwise, the values are displayed for the selected tag. Next you will enter values for the finishes.

13. In the *Name* field, enter **BEDROOM**, so the three rooms are labeled office.

14. Enter the following for the finishes (Figure 12-3.5):
    a. Base Finish: **Wood**
    b. Ceiling Finish: **GB** *(Gb = gypsum board)*
    c. Wall Finish: **PAINT-1**
    d. Floor Finish: **CARPET-1**

FYI:
When the Room Tag is selected, you can change its type from the Type Selector. The other option available, by default, is one without the square footage listed below the room number. Also, like the door schedule, a room finish schedule was already created in the template you started with!

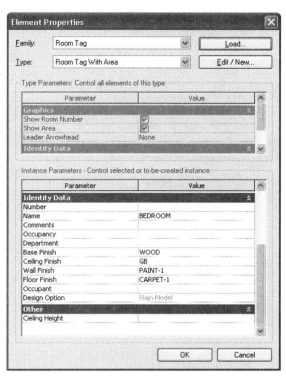

Figure 12-3.5 Element Properties – Room Tags

15. Click **OK**.

16. Switch back to the **Room Schedule** view to see the automatic updates. (Figure 12-3.6)

You can also enter data directly into the Room Schedule view.

17. Enter the following data for the Men's and Women's toilet rooms:
    a. Base: **COVED CT**
    b. Ceiling: **Gyp. Bd.**
    c. Wall: **CT**
    d. Floor: **CT**

Hopefully, in the near future, Revit will be able to enter the finishes (material) based on the wall, floor and ceiling type previously created!

TIP:

You can add fields and adjust formatting anytime by right-clicking on the schedule view and selecting View Properties. This gives you the same options that were available when you created the schedule.

			Room Schedule			
Number	Name	Base Finish	Floor Finish	Wall Finish	Ceiling Finish	Area
100	ENTRY					282 SF
101	COAT CLOSET					39 SF
102	LIVING ROOM					786 SF
103	KITCHEN					280 SF
104	MUD ROOM					87 SF
105	BATH					38 SF
106	DINING					312 SF
107	GARAGE					749 SF
200	HALLWAY					345 SF
201	OFFICE					241 SF
202	BEDROOM	WOOD	CARPET-1	PAINT-1	GB	343 SF
203	Room					44 SF
205	Room					81 SF
206	BEDROOM	WOOD	CARPET-1	PAINT-1	GB	207 SF
207	BEDROOM	WOOD	CARPET-1	PAINT-1	GB	272 SF
204	Room					41 SF
208	Room					Not Tagged
001	HALLWAY					276 SF
002	Room					39 SF
003	FAMILY ROOM					754 SF
004	SHOP					271 SF
005	MECH RM					80 SF
006	Room					36 SF
007	CRAFT ROOM					298 SF

Figure 12-3.6 Room Schedule. with new data

18. **Save** your project as **ex12-3.rvt**.

Self-Exam:
The following questions can be used as a way to check your knowledge of this lesson. The answers can be found at the bottom of this page.

1. Doors are automatically numbered based on the room number. (T/F)

2. The area for a room is calculated when a room tag is placed. (T/F)

3. Revit can tag all the doors not currently tagged on a given level with the "tag all not tagged" tool. (T/F)

4. You can add or remove various fields in a door or room schedule. (T/F)

5. Click Delete on the Options Bar to delete a door from a schedule. (T/F)

Review Questions:
The following questions may be assigned by your instructor as a way to assess your knowledge of this section. Your instructor has the answers to the review questions.

1. You can add a door tag with a leader. (T/F)

2. You can add a fire rating by selecting a door, then Properties. (T/F)

3. A door can be deleted from the door schedule. (T/F)

4. The schedule formatting only shows up when you place the schedule on a plot sheet. (T/F)

5. It is not possible to add the finish information (i.e., base finish, wall finish) to multiple rooms at one time. (T/F)

6. To modify the schedule view, right-click and then View Properties. (T/F)

7. Use the _____ dialog to adjust the various fields associated with each room tag in a plan view.

8. Most door schedules are sorted by the _____ field.

9. You can select a different Room Tag via the Type Selector. (T/F)

10. A schedule is just another way of looking at the building information model; a tabular view of the project database. (T/F)

Self-Exam Answers:
1 - F, **2** - T, **3** - T, **4** - T, **5** - T

Lesson 13
Residence: Photo-Realistic Rendering::

You will take a look at Revit's photo-realistic rendering abilities. Rather than reinventing the wheel, Revit chose to use an established architectural rendering program called Accurender.

Exercise 13-1:
Creating an exterior rendering

The first thing you will do is set up a view. You will use the camera tool to do this. This becomes a saved view that can be opened at any time from the *Project Browser*.

Creating a Camera view:

1. Open the **First Floor** plan view and **Zoom All to Fit**, so you can see the entire plan.

2. From the *View* tab, select **Camera**.

3. Click the mouse in the lower right corner of the screen to indicate the camera eye location. ***NOTICE****: Before you click, Revit tells you it wants the eye location first on the Status Bar*.

4. Next click near the entry doors; see Figure 13-1.1.

Revit will automatically open a view window for the new camera. Take a minute to look at the view and make a mental note of what you see and don't see in the view (Figure 13-1.2).

5. Switch back to the **First Floor** plan view.

6. Adjust the camera, using its grips, to look similar to Figure 13-1.3. ***TIP****: If the camera is not visible in plan view, right click on the 3D view name in the Project Browser (3D View 1) and select Show Camera*.

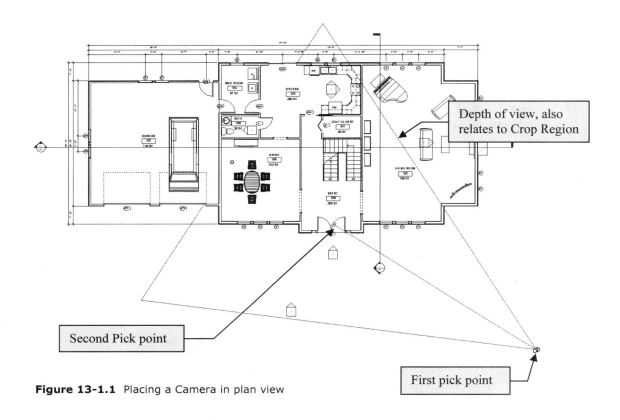

Depth of view, also relates to Crop Region

Second Pick point

First pick point

Figure 13-1.1 Placing a Camera in plan view

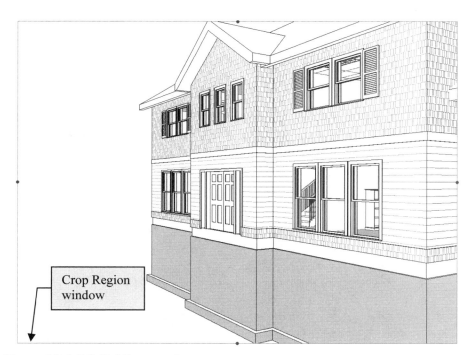

Crop Region window

Figure 13-1.2 Initial Camera view

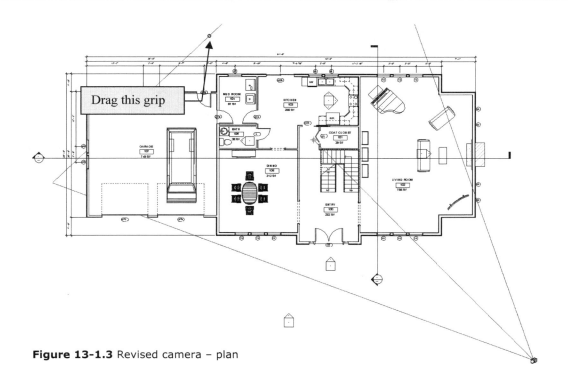

Figure 13-1.3 Revised camera – plan

7. Now switch to **3D View 1** and adjust the **Crop Region** to look similar to **Figure 13-1.4**.

This will be the view you render later in this exercise.

Figure 13-1.4 Revised camera – 3D View 1

Assigning materials to objects:

Materials are scanned images or computer generated representations of the materials your building will be made of.

Typically materials are added while the project is being modeled. For example, when you create a material (using the *Materials...* command under the *Settings* menu), you can assign an Accurender material at that time. Of course, you can go back and add or change it later. Next you will change the material assigned for the brick chimney.

8. Switch to **First Floor** plan view.

9. Select the fireplace.

10. Click **Properties** from the *Options Bar*.

11. Click **Edit/New**.

12. Click on the **Exterior Material** *value,* notice the material selected for the exterior finish is **Masonry – Brick**. (Figure 13-1.5)

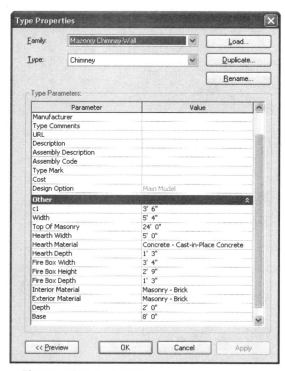

Figure 13-1.5 Chimney properties

13. Click <u>down-arrow</u> icon to the right of the label **Masonry – Brick**.

Now you will take a look that the definition of the material Masonry – Brick.

*FYI: From the Settings menu you can also select **Materials...***

You are now in the *Materials* dialog; you can also get here from the *Settings* pull-down menu. You should notice that a material is already selected. Next you will select a different brick material.

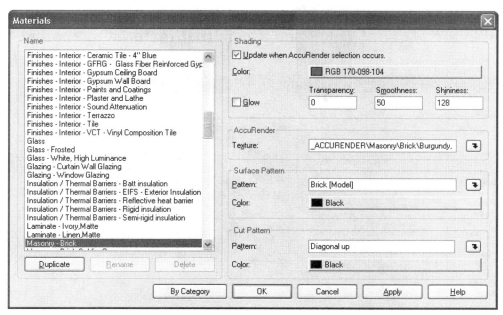

Figure 13-1.6 Materials dialog

14. In the **AccuRender** area, click the <u>down-arrow</u> to the right.

You will now see Accurender's *Material Library* dialog. The current material will be selected and displayed. (Figure 13-1.7)

You can browse through the folders on the left and select any material in the *Name* list to be assigned to the *Masonry – Brick* material in Revit. The material does not have to be brick but would be confusing if something else where assigned to the *Masonry – Brick* material name.

15. Scroll down in *Brick* category and select **Brown,_8",Running**, and then click **OK**.

Notice the Texture listed in the Accurender area is now updated.

16. Click **OK** to close the *Materials* dialog.

Now, when you render, any object (wall, ceiling, etc.) that has the material "Masonry – Brick" associated with it, will have the Broen brick on them.

If you need more than one brick color, you simply create a new material in the *Material* dialog and assign that material to another element.

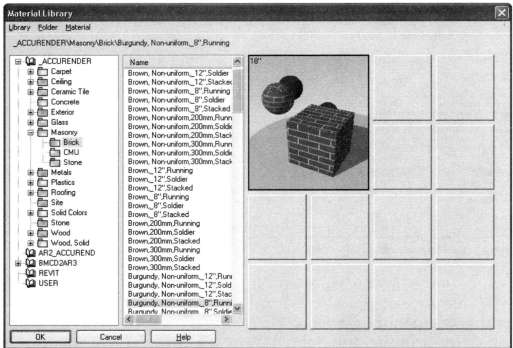

Figure 13-1.7 Accurender's Material Library dialog

Setting up the Environment:

You have several options for setting up the building's environment. You can adjust the sun settings, ground plane and sky, to name a few. You will review these options next.

17. Switch to your camera view: **3D View 1**.

18. Select the **Rendering** tab. *FYI: Most of the tools are grayed-out if you are not in a 3D view.*

19. Click on the **Settings** tool. (Figure 13-1.8)

20. Click **Exterior Scene** at the bottom and select **OK** to continue. (Figure 13-1.9)

Figure 13-1.8
Rendering Tab

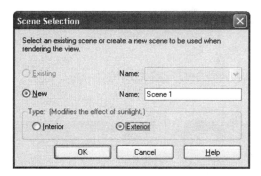

Figure 13-1.9 Scene Selection dialog

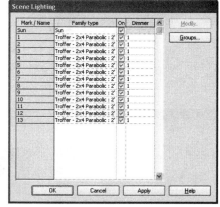

Figure 13-1.10 Scene Lighting dialog

21. Click on the **Lighting** button. (Figure 13-1.11)

You will now see a dialog similar to the one shown above (Figure 13-1.10).

You will see a *Sun* and several 2x4 light fixtures. The light fixtures relate to the fixtures you inserted in the reflected ceiling plans. It is very convenient that you can place lights in the ceiling plan and have them ready to render whenever you need to.

22. Click **Cancel** to close the *Scene Lighting* dialog.

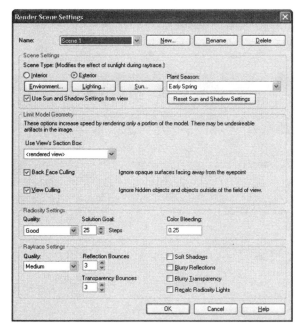

Figure 13-1.11 Render Scene Settings

23. **Uncheck** Use *Sun and Shadow Settings from view.*

24. Click on the **Sun** button.

25. On the *Date and Time* tab set (Figure 13-1.12). **TIP:** *Make sure the "specify solar angle" is set to <u>By Time, Date and Place</u>.*
    a.   Month: **6**
    b.   Day: **30**
    c.   Time: **3:00pm**

26. On the *Place* tab set (Figure 13-1.12):
    a.   Maps: *your area*
    b.   Pick *your city from the list or click your location on the map.*

27. On the *Settings* tab set the <u>Cloudiness</u> to **0.20**. (Figure 13-1.12)

28. Click **OK** to close the *Sun and Sky Settings* dialog.

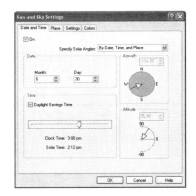  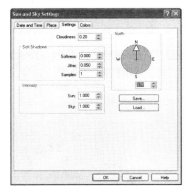

Figure 13-1.12 Accurender's Sun settings dialog (3 images)

29. Click on the **Environment** button. (Figure 13-1.11)

30. Check the **Ground Plane** option. *Notice a Ground Plane tab is added.*

31. When the *Ground Plane* tab opens, click on the **Material...** button.

32. In the *Accurender → Site* folder, select **Grass, Rye, Dark** and then click **OK** two times to close the dialog boxes.

33. Back in the *Render Scene Settings* dialog, make sure **Exterior Scene** is selected and adjust the **Raytrace Quality** to *Good*.

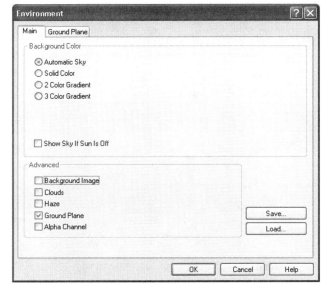

34. Click **OK**.

Next you will place a few trees into your rendering. You will adjust their exact location so they are near the edge of the framed rendering, so as not to cover too much of the building.

Figure 13-1.13 Accurender's Environment dialog

35. Switch to the First Floor plan view and select **Component** from the *Design Bar*.

36. Pick **Tree – Deciduous : Betula Pendula 19'** from the *Type Selector* on the *Design Bar*. **FYI**: *If the tree is not listed in the type selector, click Load from Library and load the Deciduous tree family from the Plantings folder.*

37. Place four trees as shown in **Figure 13-1.14** (you will make one smaller in a moment).

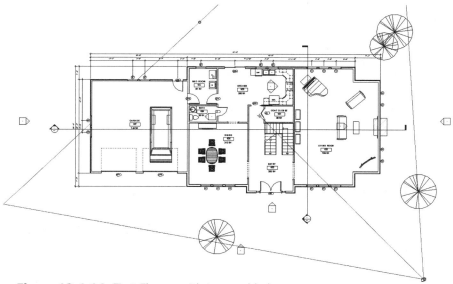

Figure 13-1.14 First Floor – with trees added

38. Adjust the trees in plan view, reviewing the effects in the *3D View 1* view, so your 3D view is similar to Figure 13-1.15. **TIP**: *Adjust the crop region if required.*

39. In the First Floor plan view, select the tree that is shown smaller in **Figure 13-14**.

40. Select *Properties* on the *Option Bar*, and then click *Edit/New*. Click **Duplicate** and enter the name: **Tree – Deciduous : Betula Pendula 15'**.

41. Change the *Plant Height* to **15'** (was 19') and then click **OK** to close the open dialog boxes.

The previous three steps allow you to have a little more variety in the trees being placed. Otherwise, they would all be the same height, which is not very natural.

Figure 13-1.15 3D View 1 – with trees

42. Open the **3D View 1** camera view.

43. Click on the **Image Size** button from the *Rendering* Tab. (Figure 13-1.8)

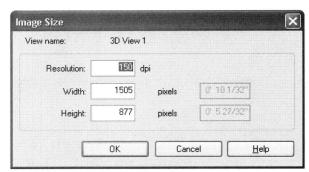

44. Change the DPI from 72 to **150**; click **OK**. (Figure 13-1.16)
    FYI: *This increases the quality of the image and how big you can print the image.*

Figure 13-1.16 Image Size

45. Click **Raytrace...** from the *Rendering* Tab. (Figure 13-1.8)

46. Click the green **GO** button on the *Options Bar*.

47. Click **NO** to the warning about lights being on. (Figure 13-1.17)

Figure 13-1.17 Lighting warning

After a few minutes, depending on the speed of your computer, you should have a rendered image similar to Figure 13-1.18 below. You can increase the quality of the image by adjusting the quality setting in the *Settings* dialog and by increasing the DPI resolution in the *Image Size* dialog. However, these higher setting require substantially more time to generate the rendering. The last step before saving the Revit project file is to save the rendered image to a file. *FYI, each time you make changes to the model (that are visible from that view), you will have to re-render the view to get an updated image.*

Depending on exactly how your view was setup, you may be able to see light from one of the light figures in the second floor office. Also, notice the stair / railing through the window' Revit has the glazing in windows set to be transparent!

Figure 13-1.18 Rendered view

48. From the *Rendering* tab select **Export Image**.
    FYI: The Capture Rendering tool saves the image within the Revit Project for placement on Sheets, this is convenient but does make the project size larger so you should delete old ones.

49. Select a *location* and provide a *file name*.

50. Set the *Save As* type: to **JPEG**.

51. Click **Save**.

The image file you just saved can now be inserted into MS Word or Adobe Photoshop for editing.

Don't forget about your Design Options:

You can render your design options in addition to the primary options. As mentioned before, you can copy a View and change its Visibility settings to show specific design options or you can just change the settings for the current view. You will do that next.

52. In the **3D View 1** view, adjust the **View Properties** so the curved entry roof and taller entry windows are visibly. (Refer back to the Design Options section if necessary.)

53. Switch to plan view and change the tree in front of the house to another option available from the *Type Selector* or the Library.

54. **Render** the image. (Figure 13-1.19)

55. **Save** your project as **ex13-1.rvt**.

Figure 13-1.19 Rendered view

Notice the large tree casts shadows on the ground and the building, you can also see reflections of the tree in the windows. Remember, the angle and location of the shadows are based on the time, date and location on the earth; if you change those settings, the shadows will change accordingly! Also notice the ground plane that was added and the fact that it extends to the horizon.

As you can imagine, an image like this could really "sell" your design to the client. Revit makes the process to get to this point very easy compared to legacy CAD programs and techniques used prior to Revit.

Exercise 13-2:
Rendering an isometric in section

This exercise will introduce you to a view tool called *Section Box*. This tool is not necessarily related to renderings, but the two tools together can produce some interesting results.

Setting up the 3D view:

1. Open file ex13-1.rvt and **Save As ex13-2.rvt**.

2. Switch to the **3D** view via the icon on the *View* toolbar *(not the 3D View 1 from exercise 13-1)*.

3. *Right-click* in the drawing area and select **View Properties**.

4. Activate the **Section Box** parameter and then click **OK**.

You should see a box appear around your building, similar to Figure 13-2.1. When selected, you can adjust the size of the box with its grips. Anything outside the box is not visible. This is a great way to study a particular area of your building while in an isometric view. You will experiment with this feature next.

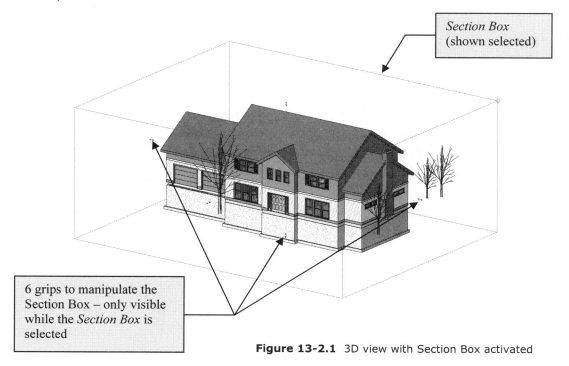

Section Box
(shown selected)

6 grips to manipulate the
Section Box – only visible
while the *Section Box* is
selected

Figure 13-2.1 3D view with Section Box activated

5. To practice using the **Section Box**, drag the grips around until your view looks similar to **Figure 13-2.2**. *TIP: This may require the Dynamically Modify View tool as well or hold Shift + drag Wheel button.*

    a. Click and drag the grip for the top downward.
    b. Click and drag the grip for the east (right) face toward the west (left).

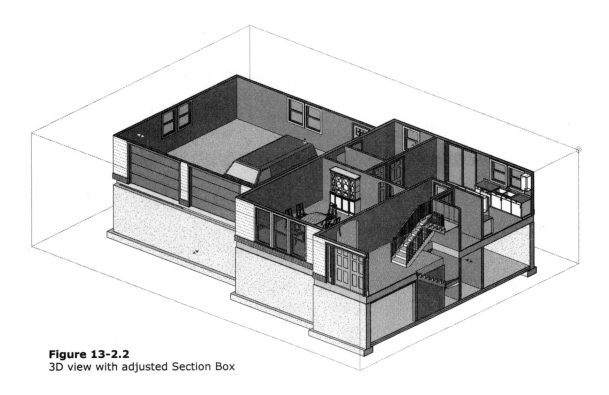

Figure 13-2.2
3D view with adjusted Section Box

Did you notice the grips look different? One grip looks like two arrows rather than a small square. You can still click on any part of the grip and drag it, but you can also single-click on one of the arrows to slightly adjust the section box in the direction the arrow points. This is similar to the "nudge" feature; when an element is selected and you use the arrows keys to move it just a little bit.

This creates a very interesting view of the First Floor plan. What client would have trouble understanding this drawing?

6. Now re-adjust the **Section Box** to look similar to **Figure 13-2.3**.

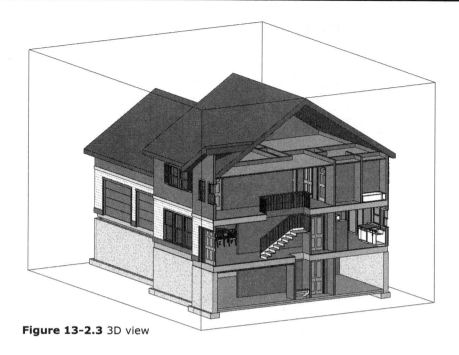

Figure 13-2.3 3D view

7. From the *Rendering* tab, select **Settings** and change the following *Sun* settings (Figure 13-2.4):
   a. Month: **2** (February)
   b. Day: **28**
   c. Time: **8:28am**

FYI:
Don't forget, without doing any rendering in Revit you can turn on the real-time shadows in any view. You can even turn on shadows in your 2D exterior elevation view (see below) for a great effect will little effort!

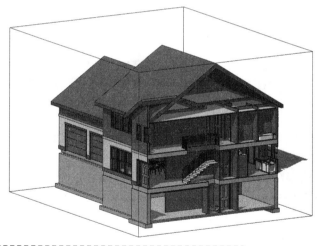

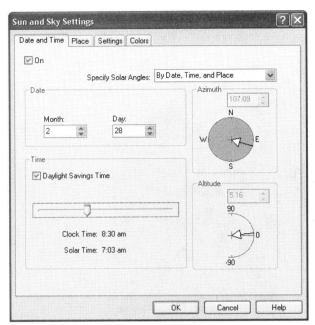

Figure 13-2.4 Modified Sun settings

8.   Select **Raytrace** and then the green **Go** button on the *Options Bar*. You will get a warning stating the rending will take long if the *Region Raytrace* is not used. Click **Cancel**.

9.   Select **Region Raytrace** and select a window around the entire building and then click **GO**. *TIP: This tool is nice for checking a material before rendering the entire building, which takes longer.*

The image will take a few minutes to render (again, depending on the speed of your computer). When finished it should look similar to **Figure 13-2.5**. Notice that, if an object does not have a material associated with it, the object is rendered using the "shade mode" color. The image looks much better on the screen or printed in color.

Both the "shade mode" and "rendered" views have their pluses and minuses. The "shade mode" has nice crisp lines defining all the edges, and a "rendered" image has shadows from the Sun (however, now in version 7 you have the real-time shadows).

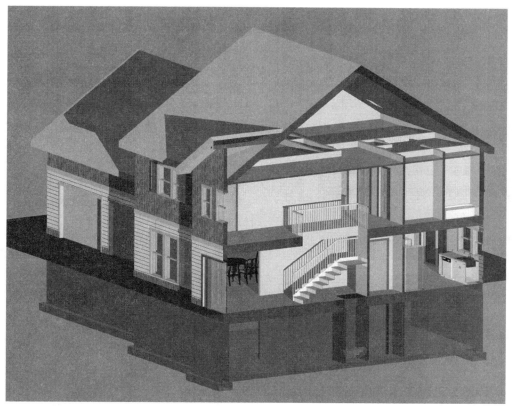

Figure 13-2.5 Rendered isometric view

Adjusting an object's material:

As previously mentioned, most objects already have a material assigned to them. This is great because it allows you to quickly render your project to get some preliminary images. However, they usually need to be adjusted. You will do this next.

10. 3D view, on the *View Control Bar*, change the shade mode to **Shaded with Edges On**; the rendered view disappears.

11. Select one of the cabinets in the kitchen.

12. Click the **Properties** button from the *Options Bar*.

13. Click the **Edit/Copy**... button to set the *Type Properties*.

Notice the Cabinet Material parameter is set to **<By Category>** (Fig. 13-2.6). You could change the material here for just the selected cabinet; however, by setting a material to be *By Category* you can quickly change the material for all the cabinets at one time (without having to select every cabinet).

14. **Close** the dialogs.

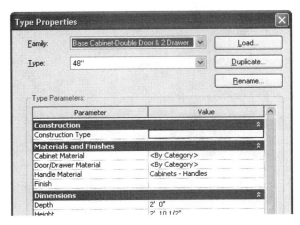

Figure 13-2.6 Cabinet properties

15. Right-click on the drawing window and select **View Properties**.

16. Click **Edit** next to the *Visibility* parameter.

17. On the *Model Categories* tab, select **Object Styles...** button.

18. Expand the **Casework** category. (Fig. 13-2.7)

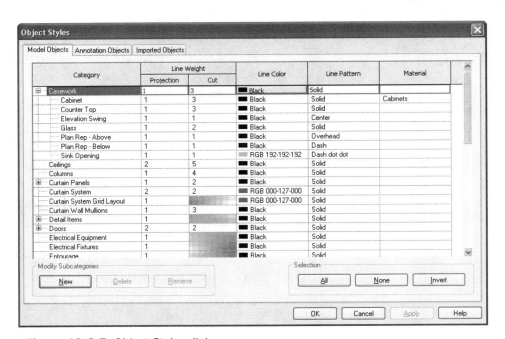

Figure 13-2.7 Object Styles dialog

Notice that *Cabinet* has a material assigned to it called **Cabinets**.

19. Click on **Cabinets** in the *Material* column, and then click the down-arrow button that appears to the right.

You are now in the *Materials* dialog box, where you can create and edit *Materials*.

20. Select **Cabinets** from the *Name* list (it should be selected).

21. Change the Accurender material to: **\Wood \ Cherry \ Natural, Low Gloss**. (Figure 13-2.8)

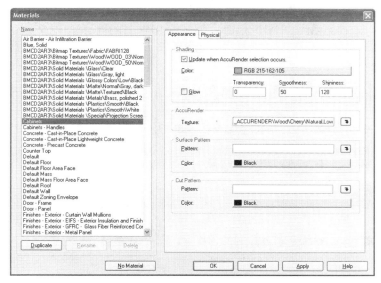

Figure 13-2.8 Accurender material selector

22. Click **OK** to close the open dialog boxes.

23. You can now re-render the 3D view and see the results.

24. **Save** your project as ex **13-2.rvt**.

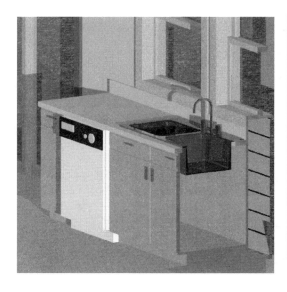

TIP:
While in the 3D view with the Section Box active, you can zoom in to an area (just like you can in other views) and render that enlarged area. The larger area will have more detail; for example, you can see the wood grain on the cabinets.

Similarly, you can change the material for the counter top and dishwasher, etc.

Exercise 13-3:
Creating an interior rendering

Creating an interior rendering is very similar to an exterior rendering. This exercise will walk through the steps involved in defining exterior light sources and using radiosity to create high quality interior renderings.

Setting up the camera view:

1. Open ex13-2.rvt and **Save As ex13-3.rvt**.

2. Open **First Floor** plan view.

3. From the *View* tab, select **Camera**.

4. Place the *Camera* as shown in **Figure 13-3.1**.

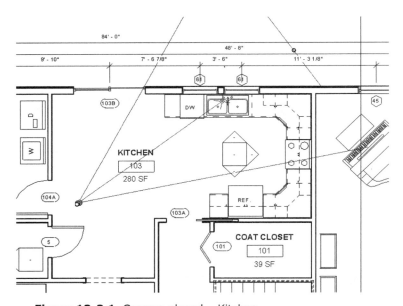

Figure 13-3.1 Camera placed – Kitchen

Revit uses default heights for the camera and the target. These heights are based on the current level's floor elevation. These reference points can be edited via the camera properties.

Revit will automatically open the newly generated camera view. Your view should look similar to **Figure 13-3.2**. *FYI: Make sure you created the camera on Level 2 and picked the points in the correct order.*

Figure 13-3.2 Initial interior camera view

5.   Using the **Crop Region** rectangle, modify the view to look like **Figure 13-3.3**.
6.   Copy the two trees in the back of the house, into a position so they can be seen through the window over the sink.
     TIP: You will have to switch to plan view to adjust the camera's depth of view to see the trees. Reminder: if the camera does not show in plan view, right-click on the camera view label (while in plan view) in the project browser and select Show Camera. You may need to switch between the plan and camera view several times to adjust the trees so they are visible through the windows.

7.   Switch back to **First Floor** plan to see the revised *Camera* view settings.

Notice the field of view triangle is wider based on the changes to the Crop Region in the camera view. (Figure 13-3.4)

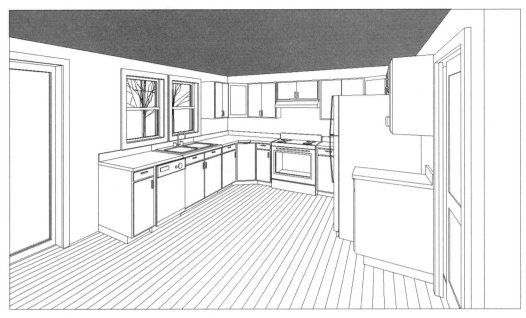

Figure 13-3.3 Modified interior camera

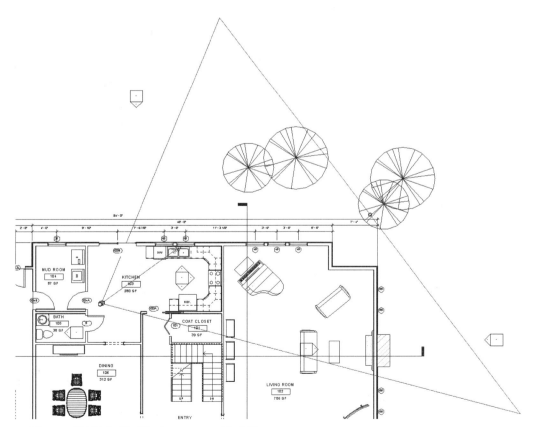

Figure 13-3.4 Modified camera – First Floor

8. Select the *Camera* and click the **Properties** button on the *Option Tab*. (Figure 13-3.5)

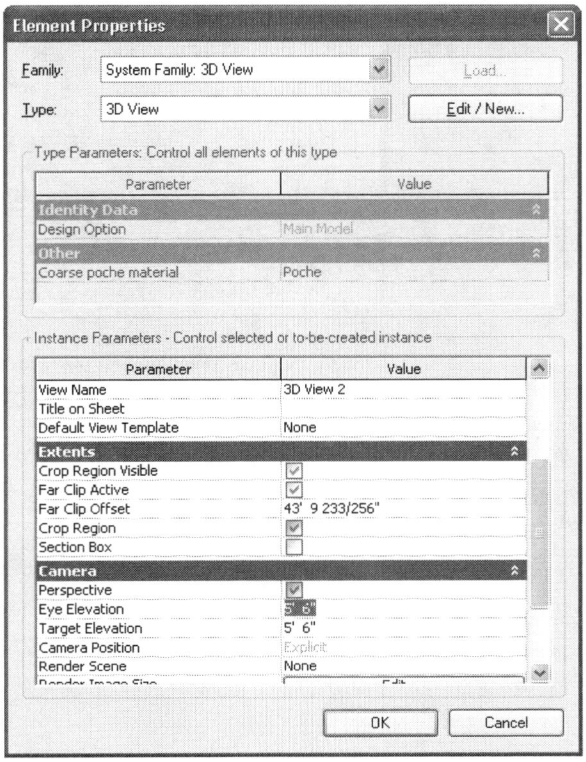

Figure 13-3.5 Camera properties

9. Make sure the **Eye Elevation** is **5'-6"**.

Notice the other settings for the camera, changing the Target Elevation would cause the camera to look up (at the ceiling), or down (at the floor).

10. Click **OK**.

The vertical lines are distorted due to the wide field of view (crop region). This is similar to what a camera with a 10-15mm lens would get in the finished building.

Setting up day lighting for radiosity:

Radiosity is a method of calculating natural light for an interior scene. You need to define the exterior light source. Revit (via Accurender) then calculates how the light bounces off different surfaces. This creates a very realistic interior rendering.

11. Switch to the **Default 3D** view and turn off *Section Box*.

12. From the *Rendering* tab, select **Daylights...**

13. Select the two windows above the sink; when one is selected it is shown red while in the *Daylights* tool. You need to select near the perimeter of the element to select it. Also select the sliding glass door. When finished, all the glass will have a red tint.

14. Click the **Modify** button to exit the *Daylights* tool.

Next you will adjust the view Setting for Rendering.

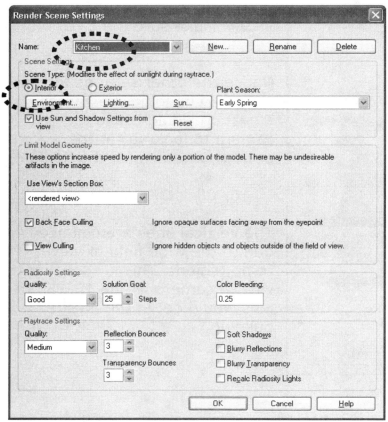

Figure 13-3.6 Settings

15. Switch to the camera view (3D View 2).

16. Click **Settings** from the *Rendering* tab.

17. Click *New* to create a scene named **Kitchen**.

18. Select *Interior Scene* and then **OK**. (Figure 13-3.6)

Next you will run the *Radiosity* tool to calculate the lighting before the *Raytrace* tool is used.

19. Select **Radiate...** from the *Rendering Tab*.

20. Click **OK** to continue. (Figure 13-3.7)

Revit will take a few minutes to *Radiate* the current view. When finished, your view should look similar to **Figure 13-3.8**.

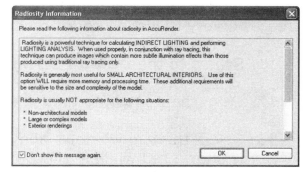

Figure 13-3.7 Warning

Figure 13-3.8

Next you will render the view.

    21. Select **Raytrace...** from the *Rendering* tab.

This will take several minutes depending on the speed of your computer. When finished, the view should look similar to **Figure 13-3.9**.

    22. Click **Export Image** from the Rendering tab to save the image to a file on your hard drive. Name the file **Kitchen.jpg** (jpeg file format).

You can now open the *Interior Atrium.jpg* file in Adobe Photoshop or insert into MS Word type program to manipulate or print.

To toggle back to the normal hidden view, **Display Model** from the *Rendering* tab.

There are many things you can do to make the rendering look even better. You can add interior light fixtures and props (e.g. pictures on the wall, items on the counter top, and lawn furniture in the yard). Once you add interior lights, you can adjust the Sun setting to night time and then render a night scene.

Figure 13-3.9 Rendered view

Now that you have the Scene setup and the Radiosity calculated for the kitchen, you can quickly add another Camera to look at the other half of the kitchen; rendering it using the "kitchen" scene settings setup for the previous kitchen camera view. The ability to stretch the Crop Region allows you to create interesting view like the one below, which is really more like a panoramic view than a traditional camera view (Figure 13-3.10).

Figure 13-3.10 Another rendered view

23. **Save** your project as **13-3.rvt**.

Exercise 13-4:
Adding people to the rendering

Revit provides a few RPC people to add to your renderings. These are files from a popular company that provides 3D photo content for use in renderings (http://www.archvision.com). You can buy this content in groupings (like college students) or per item. In addition to people, they offer items like cars, plants, trees, office equipment, etc.

Loading content into the current project

1. Open ex13-3.rvt and **Save As 13-4.rvt**.

2. Switch to the **First Floor** plan view.

3. Select the **Component** tool from the *Rendering* tab.

4. Click the **Load...** button on the *Options Bar*.

5. Browse to the **Entourage** folder and select both the **RPC Male** and **RPC Female** files (using the Ctrl key to select both at once) and click **Load**.

6. Place one **Male** and one **Female** as shown in **Figure 13-4.1**.

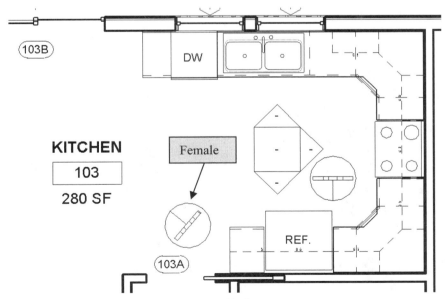

Figure 13-4.1 Kitchen – RPC people added

The line in the circle (Figure 13-4.1) represents the direction a person is looking. You simply *Rotate* the object to make adjustments. The image below shows the RPC people that can be placed in your project.

Figure 13-4.2 Type Selector – Options Bars

7. Switch to your interior kitchen camera view.

8. Select **Settings** on the *Rendering* tab.

9. Set the *Scene* to **Kitchen**.

10. Click **OK**.

11. Click **Raytace** to render the view with the people in the view.

Figure 13-4.3 Interior kitchen view with people added

Your rendering should now have people in it and look similar to **Figure 13-4.3**. Notice how the people cast shadows; remember the only light source is the Sun thought the two windows and the glass door. If you want to change the people, simply select them and pick from the *Type Selector*.

Adding people and other "props" gives your model a sense of scale and makes it look a little more realistic. After all, architecture is for people. These objects can be viewed from any angle. Try a new camera view from the other angle to see how the people adjust to match the view and perspective, see Figure 13-4.4.

Try adding one of the RPC vehicles to the exterior and render the exterior of the building again; you would not want to use Radiosity on the exterior scene due to the amount of time it would add to the render time.

12. **Save** your project as **ex13-4.rvt**.

> **FYI:**
> As with other families and components, the more you add to your project, the bigger your project file becomes. It is a good idea to load only the items you need and delete the unused items via the Project Browser. Your project should be about 10.5MB at this point in the tutorial.

Self-Exam:
The following questions can be used as a way to check your knowledge of this lesson. The answers can be found at the bottom of this page.

1. Creating a camera adds a view to the Project Browser list. (T/F)

2. Accurender materials are defined in Revit's Materials dialog box. (T/F)

3. After inserting a light fixture, you need to adjust several setting before rendering in Accurender and getting light from the fixture. (T/F)

4. You can adjust the season, which affects how the trees are rendered. (T/F)

5. Use the _____ _____ tool to hide a large portion of the model.

Review Questions:
The following questions may be assigned by your instructor as a way to assess your knowledge of this section. Your instructor has the answers to the review questions.

1. You cannot pick a material for the ground plane. (T/F)

2. Radiosity is best used on exterior scenes. (T/F)

3. Adding components and families to your project does not make the project file bigger. (T/F)

4. Using the Radiosity feature adds a significant amount of time to the rendering process. (T/F)

5. The RPC people can only be viewed from one angle. (T/F)

6. You have to use the _____ tool (to define the exterior light source) before using the Radiate tool.

7. Adjust the _____ _____ to make more of a perspective view visible.

8. You use the _____ tool to load and insert RPC people.

9. You can adjust the Eye Elevation of the camera via the camera's

_____.

10. What is the file size of (completed) exercise 13-4? _____ MB

Lesson 14
Residence: Construction Documents Set::

This lesson will look at bringing everything you have drawn thus far together onto sheets. The sheets, once set up, are ready for plotting. Basically, you place the various views you have created on sheets. The scale for each view is based on the scale you set while drawing that view (which is important to have set correctly because it affects the text and symbol sizes). When finished setting up the sheets, you will have a set of drawings ready to print, individually or all at once.

Exercise 14-1:
Setting up a sheet

Creating a Sheet view:

1. Open ex13-4.rvt and **Save As** 14-1.rvt.

2. Select **Sheet...** from the *View* tab.   Sheet...

Next Revit will prompt you for a Titleblock to use. The template file you started with has two: 11 x 17 and 22 x 34. (Figure 14-1.1)

3. Select the **D 22x34 Horizontal** titleblock and click **OK**.

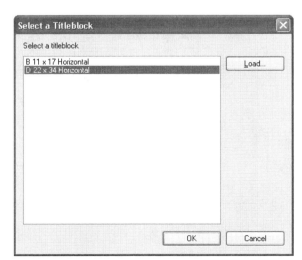

> **NOTICE:**
> A new view shows up in the Project Browser under the heading *Sheets*. Once you get an entire CD set ready, this list can be very long.

Figure 14-1.1 Select a Titleblock

Figure 14-1.2 Initial Titleblock view

4. **Zoom** into the sheet number area (lower right corner).

5. Adjust the text to look similar to **Figure 14-1.3**; click on the text and edit (make sure *Modify* is selected).

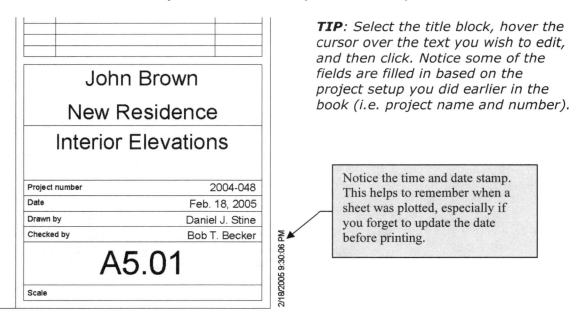

TIP: Select the title block, hover the cursor over the text you wish to edit, and then click. Notice some of the fields are filled in based on the project setup you did earlier in the book (i.e. project name and number).

Notice the time and date stamp. This helps to remember when a sheet was plotted, especially if you forget to update the date before printing.

Figure 14-1.3 Revised Titleblock data

6. **Zoom out** so you can see the entire sheet.

7. With the sheet fully visible, click and drag the **Living Room - East** label (under Elevations) from the *Project Browser* onto the sheet view.

You will see a red box that represents the extents of the view you are placing on the current sheet.

8. Move the *Viewport* around until the box is in the upper-right corner of the sheet (this can be adjusted later at any time).

Your view should look similar to **Figure 14-1.4**.

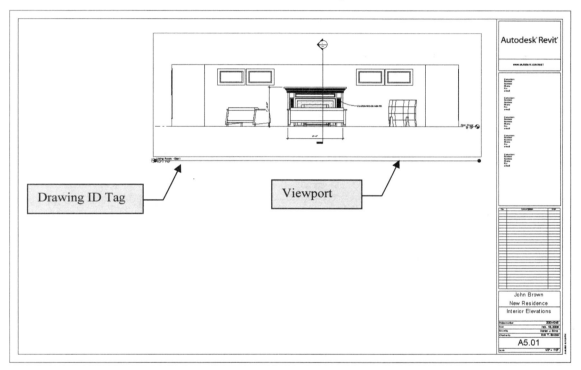

Figure 14-1.4 Sheet with Living room view

9. Click the mouse in a "white" area (not on any lines) to deselect the Living Room - East view. Notice the red box goes away.

10. **Zoom In** on the lower left corner to view the drawing identification symbol that Revit automatically added. (Figure 14-1.5)

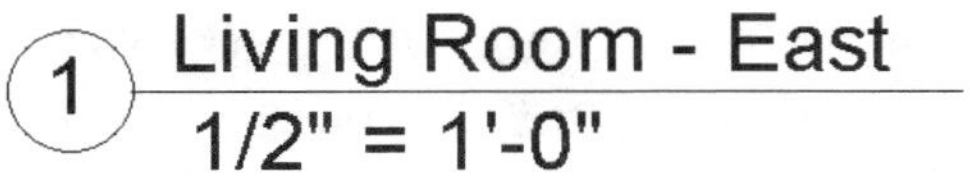

$$\begin{array}{c} \text{(1)} \end{array} \frac{\text{Living Room - East}}{1/2" = 1'-0"}$$

Figure 14-1.5 Drawing ID tag

NOTICE: *The drawing number for this sheet is added. The next drawing you add will be number 2.*

The view name is listed. This is another reason to rename the elevation and section views as you create them.

Also notice that the drawing scale is listed. Again, this comes from the scale setting for the Living Room - East view.

As you can see, in Figure 14-1.4, the furniture is showing up; normally this would be turned off. You can turn these off using the same technique covered later for the trees in the exterior elevations.

Setting up the Floor Plans:

Setting up floor plans is easy; actually, they are already setup from the template you started with.

11. From the *Sheets* section in the Project Browser, double-click on the sheet named **A1 – First Floor Plan**.

As you can see, the floor plan is already setup on a sheet. Again, a few things can be turned off (i.e. trees, RPC People) per the techniques covered next. You would also want to move the exterior elevation tags closer to the building so they are within the title block.

12. Change the sheet number to **A2.01**.

13. Close the First Floor plan *Sheet*.

Setting up the Exterior Elevations:

Next you will set up the exterior elevations. Again, the sheets are already setup but the Views have not been placed on the Sheet like the floor plans have.

14. Open Sheet **A6 – Elevations**.

15. Drag the **South** elevation view onto the sheet. Place the drawing near the bottom (centered).

Your drawing should look similar to Figure 14-1.6.

Figure 14-1.6 South exterior elevation

Next you will turn off the trees in the south view. Normally you would turn them off in all views. However, you will only turn them off in the south view to show that you can control visibility per view on a sheet.

16. Click near the edge of the drawing to select the viewport of the south elevation (reference Figure 14-1.6).

17. Now **Right-Click** and select **Activate View** from the pop-up menu.

At this point you are in the viewport and can make changes to the project model to control visibility, which is what you will do next.

18. Right-click in the "white space" and select **View Properties...**

19. Click the **Edit** button next to *Visibility*.

20. In the Visibility dialog **Uncheck Planting**.

21. Close the open dialog boxes.

22. Right-click anywhere in the drawing area and select **Deactivate View** from the pop-up menu.

Now the trees are turned off for the South Elevation, if you would have had another elevation view on this same sheet, the trees would still be visible for that view.

23. Add the **North** elevation to the **A7 – Elevations** *Sheet*.

Now you will stop for a moment and notice that Revit is automatically referencing the drawings as you place them on sheets.

24. Switch to **First Floor** (see Figure 14-1.7).

Notice in Figure 14-1.7 that the number A6 represents the sheet number that the drawing can be found on. The number one (1) is the drawing number to look for on sheet A6. The empty tag is the enlarged elevation view you setup earlier in the book; it has not been placed on a sheet yet so it is not filled in.

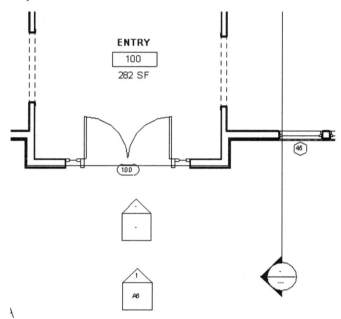

Figure 14-1.7 First Floor – elevation tag filled-in

Setting up Sections:

25. Open *Sheet* **A8 – Building Sections** and place <u>Cross Section 1</u>

26. Open *Sheet* **A9 – Building Sections** and place <u>Longitudinal Section</u> view on the *Sheet*.

27. Switch to *First Floor* plan view and zoom into the area shown in Figure 14-1.9.

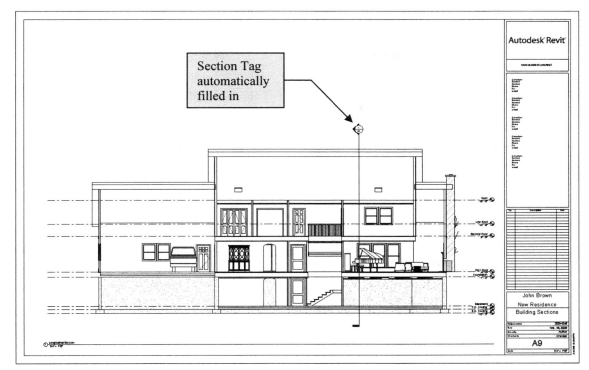

Figure 14-1.8 A9 – Building Sections sheet

Notice, again, that the reference bubbles are automatically filled in when the referenced view is placed on a sheet. If the drawing is moved to another sheet, the reference bubbles are automatically updated.

You can also see in Figure 14-1.8 (above) that the reference bubbles on the building sections are filled in.

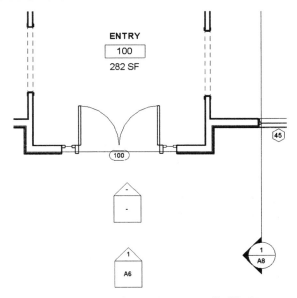

Figure 14-1.9 First Floor – Section ref's filled in

Set up the remaining sheets:

Next you set up sheets for the remaining views that have yet to be placed on a sheet (except for the 3D views).

Add the remaining views to the appropriate Sheets; if one does not exist you can create a new sheet.

Question: On a large project with hundreds of views, how do I know for sure if I have placed every view on a sheet?

Answer: Revit has a feature called *Browser Organization* that can hide all the views that have been placed on a sheet. You will try this next.

Take a general look at the *Project Browser* to see how many views are listed.

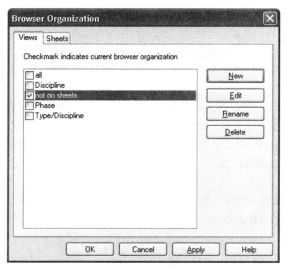

Figure 14-1.10 Browser Organization dialog

28. From the *Settings* pull-down menu select **Browser Organization...**

29. On the *Views* tab, click the check-box next to **not on sheets**. (Figure 14-1.10)

30. Click **OK**.

31. Notice the list in the *Project Browser* is now smaller. (Figure 14-1.11)

The Project Browser now only shows drawing views that have not been placed onto a sheet. Of course, you could have a few views that do not need to be placed on a sheet, but this feature will help eliminate errors. Also notice the label at the top: Views (not on sheets); this tells you what mode/filter the Project Browser is in.

Next you will reset the Project Browser.

32. Open *Browser Organization* again and check the box next to **all** and click **OK** to close the dialog box.

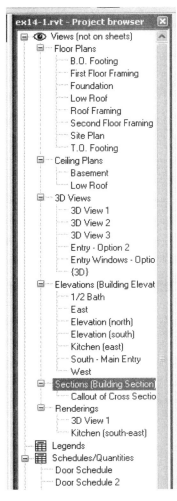

Figure 14-1.11
Project Browser; Sheets (all)

Sheets with Design Options:
Finally, you will set up a sheet to show the two Design Options.

33. Create a *Sheet* named **Entry Options** and number it **A100**.

34. Place both *Entry – Option 2 3D view's* on the new Sheet and change the scale to **1/8" = 1'-0"**.
TIP: Right-click on placed view and select Properties; change the Scale.

Figure 14-1.12
Entry Options Sheet; two views with Design Options added

Each 3D view has its Visibility modified to show the desired Design Options. When a view is placed on a sheet those settings are preserved.

35. **Save** your project as **ex14-1.rvt**.

Exercise 14-2:
Sheet Index

Revit has the ability to create a sheet index automatically. You will study this feature now.

Creating a Sheet List View

1. Open ex14-1.rvt and **Save As** ex14-2.

2. From the *View* pull-down menu select **New → Drawing List...**

You are now in the *Drawing List Properties* dialog box. Here you specify which fields you want in the sheet index and how to sort the list. (Figure 14-2.1)

3. Add **Sheet Number** and **Sheet Name** to the right. *(Click Add →)*

4. Click **OK**.

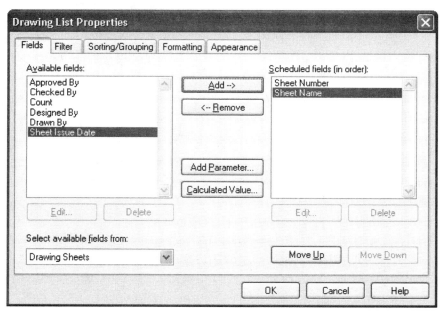

Figure 14-2.1
Drawing List Properties Dialog; sheet number and name "added"

Now you should notice that the *Sheet Names* are cut off because the column is not wide enough (14-2.2). You will adjust this next.

5. Move your cursor over the right edge of the *Drawing List* table and click-n-drag to the right until you can see the entire name (Figure 14-2.2).

Figure 14-2.2
Drawing List view; notice sheet names are cut off in right column

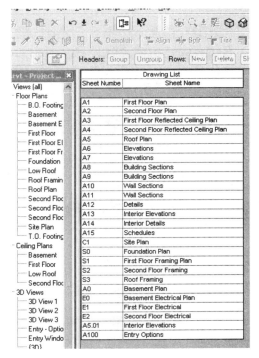

Figure 14-2.3
Drawing List view; sheet names are now visible

Adjusting which sheets show up in the list

Looking at the drawing list (Figure 14-2.3), you decide to remove the Entry options sheet from the list as it is not part of the construction document set (i.e. sheet A5.01).

As with other schedules, this is live data which is directly connected to the model. If you change a sheet number here, the number will change throughout the Project. If you delete a number (i,e, a row), Revit will delete the sheet from the project.

You will look at the option that allows you to remove a sheet from the drawing list without deleting the sheet.

6. Right-click on sheet **A100** in the drawing list and select **Delete Row(s)**. (You will NOT actually delete this sheet.)

You will get a warning stating the sheet will be deleted and what you should do if you only want to remove the sheet from the drawing list.

7. Click **Cancel**.

8. Right-click on the *Sheet* **A100 – Entry Options** in the *Project Browser* and click **Properties**.

9. Uncheck "appears" in Drawing List. (Figure 14-2.4)

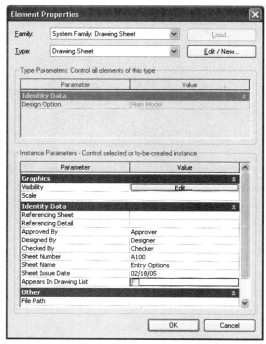

Figure 14-2.4
Properties for sheet A100

Notice the *Sheet* was removed from the *Drawing List* view.

10. In the *Drawing List*, right-click on sheet <u>A13 Interior Elevations</u> and select **Delete Row(s)** from the menu.

11. Click **OK**.

Not only was the sheet removed from the Drawing List, it was also deleted from the Project. Next you will Renumber a sheet.

12. Click in the cell with the sheet number and change the number to read A3.

Now an existing sheet has been renumbered and so has all the detail, elevation and/or section bubbles that point to this sheet.

13. Right-click on the schedule and select **View Properties**.

14. Click Edit... next to Sorting / Grouping.

15. Set *Sort By* to **Sheet Numbers** and click **OK** to close all open dialog boxes.

Now the sheets should be sorted correctly, including the sheet you just renumbered.

Setting up a Title Sheet

Now you will create a title sheet to place your sheet index on.

16. Create a new Sheet:
    a. Number: **T1**
    b. Name: **Title Sheet**

17. From the *Schedules/Quantities* category of the *Project Browser*, place (i.e. drag-n-drop) the view named **Drawing List** on the *Title Sheet*.

18. Drag the column grips so each row is only one line.

19. Create a new text style named **1" Arial**, and adjust the setting accordingly.

20. Add large text across the top of the sheet that reads "NEW RESIDENCE FOR JOHN BROWN". (Figure 14-2.5)

Next you will place one of your rendered images that you saved to file (raster image). If you have not created a raster image, you should refer back to Lesson 13 and create one now (otherwise you can use any BMP or JPG file on your hard drive if necessary).

If you used the *Capture Image* feature you can simply drag one of the images listed under *Renderings* from the *Project Browser.* The interior kitchen rendering was added to the title sheet using this method, here you can see Revit has added a Drawing Title tag beneath the image; which would allow you to reference this image from another location.

21. From the *File* pull-down menu select **Import/Link → Image...**

22. Browse to your JPG or BMP raster image file, select it and click **Open** to place the Image.

23. Click on your Title Sheet to locate the image.

Your sheet should look similar to Figure 14-2.5.

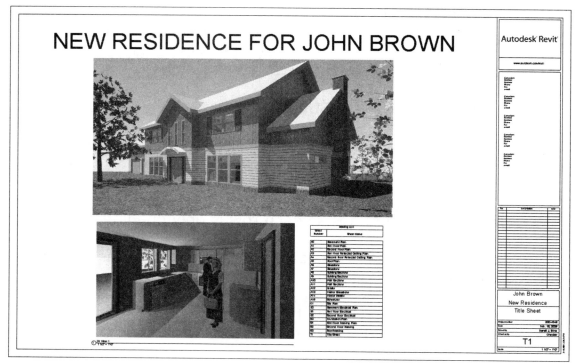

Figure 14-2.5
Sheet View: Title Sheet with drawing list, text and image added

24. From the *File* menu, select **Raster Images...**

You are now in the Raster Image dialog which gives you a little information about the image and allows you to delete it from the project (Figure 14-2.6).

FYI:
You can delete a view from a Sheet with out deleting the view from the project.

Also, you can only place a view on one sheet; you would have to duplicate the view in order to have that view repeated.

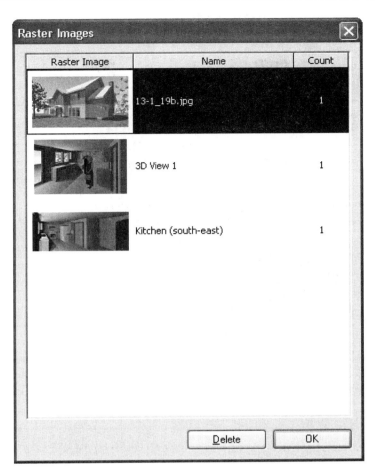

Figure 14-2.6
Raster Image dialog

25. Click **OK** to close the **Raster Images** dialog.

Exporting a 3D DWF file:

We will now take a quick look at a feature new to this release of Revit. The 3D DWF file is an easy way to share your project file with others without the need to give them you editable Revit file (i.e. your intellectual property) or the need for them to actually have Revit installed. (Side note: you can download Revit from Autodesk's website and run it in *Viewer Mode* for free; it is the full Revit software with no Save functionality). The 3D DWF file is much smaller in file size than the original Revit file. Also, the DWF *Viewer* is a small program that can be downloaded for free from www.autodesk.com.

26. Switch to your default **3D view**.

You must be in a 3D view for the 3D DWF feature to work. If you are not in a 3D view, you will get the following prompt (to right):

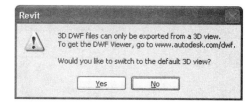

27. From the *File* menu, select **Export DWF → 3D DWF...**

28. Specify a file name and location for the DWF file; click **Save**.

That's all you need to do to create the file. Now you can email it to the client or a product rep to get a more accurate cost estimate.

The 3D DWF file is a meager 347kb, whereas the Revit Project file is 11 MB (your file sizes will vary slightly based on factors like the number of families loaded, etc.). In most cases, networks and servers are set up so that individuals cannot receive large files via email; therefore the 3D DWF is very useful.

The DWF view should have been installed with Revit. You can access it from *Start → All Programs → Autodesk → Autodesk DWF Viewer*. You can zoom, orbit and select objects. Notice the information displayed on the left when the front door is selected (see image below). Also notice, under the Windows heading, the sizes and quantities are listed.

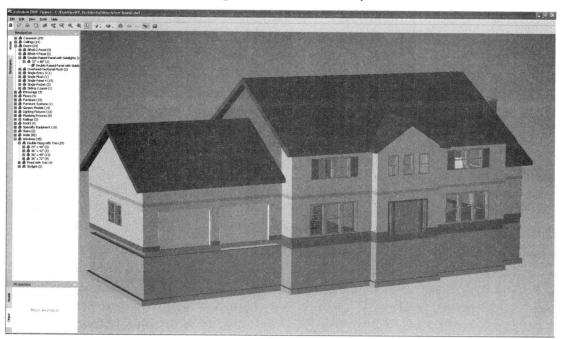

Figure 14-2.7 3D DWF in Autodesk DWF viewer

29. **Save** your project as **ex14-2.rvt**.

Exercise 14-3:
Printing a set of drawings

Revit has the ability to print an entire set of drawings, in addition to printing individual sheets. You will study this now.

Printing a set of drawings

1. **Open ex14-2.rvt**.

2. Select **Print** from the *File* pull-down menu.

3. In the *Print range* area, click the option **Selected views/ sheets**. (Figure 14-3.1)

4. Click the **Select...** button within the *Print range* area.

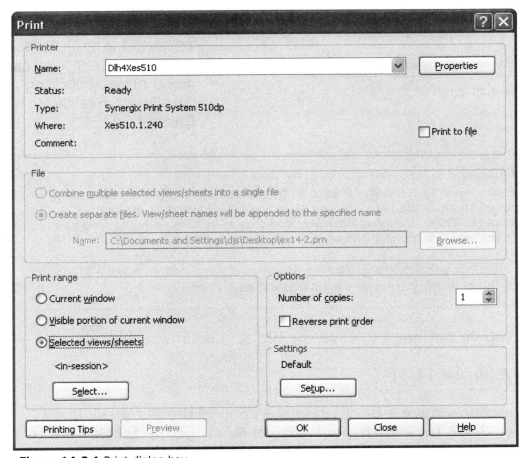

Figure 14-3.1 Print dialog box

You should now see a listing of all *Views* and *Sheets*. (Figure 14-3.2)

Notice at the bottom you can show both Sheets and Views, or each separately. Because you are printing a set of drawing you will want to see only the sheets.

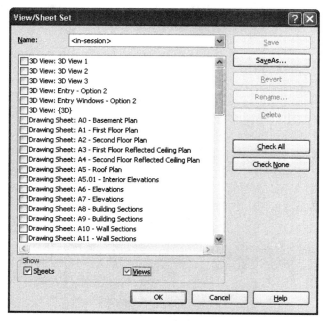

Figure 14-3.2 Set tool for printing

5. **Uncheck** the **Views** option.

The list is now limited to just *Sheets* that have been set up in your project.

6. Select all the Drawing Sheets except A100.

7. Click **OK** to close the **View/Sheet Set** dialog.

8. IF YOU ACTUALLY WANT TO PRINT A FULL SET OF DRAWINGS, you can do so now by clicking OK. Otherwise click **Cancel**.

FYI:

Once you have selected the sheets to be plotted and click OK, you are prompted to save the list. This will save the list of selected drawing to a name you choose. Then, the next time you need to print those sheets, you can select the name from the drop-down list at the top (Figure 14-3.2).

On very large projects (e.g. 20 floor plan sheets) you could have a Plans list saved, a Laboratory Interior Elevations list saved, etc.

---

9. You do not need to save the file at this time.

[End of Exercise 14-3]

You should now have a basic understanding of the Autodesk Revit software. **Gook luck with your future Revit projects!** Be sure to visit *www.augi.com* to share knowledge with other Revit users.

Self-Exam:

The following questions can be used as a way to check your knowledge of this lesson. The answers can be found at the bottom of this page.

1. You have to manually fill in the reference bubbles after setting up the sheets. (T/F)

2. You cannot control the visibility of objects per viewport. (T/F)

3. It is possible to see a listing of only the views that have not been placed on a sheet via the Project Browser. (T/F)

4. You only have to enter your name on one titleblock, not all. (T/F)

5. Use the _____ tool to create another drawing sheet.

Review Questions:

The following questions may be assigned by your instructor as a way to assess your knowledge of this section. Your instructor has the answers to the review questions.

1. You need to use a special command to edit text in the titleblock. (T/F)

2. The template you started with has two titleblocks to choose from. (T/F)

3. You only have to enter the project name on one sheet, not all. (T/F)

4. The scale of a drawing placed on a sheet is determined by the scale set in that view's properties. (T/F)

5. You can save a list of drawing sheets to be plotted. (T/F)

6. Use the _____ _____ tool to edit the model from a sheet view.

7. The reference bubbles will not automatically update if a drawing is moved to another sheet. (T/F)

8. On new sheets, the sheet number on the titleblock will increase by one from the previous sheet number. (T/F)

9. DWF files can consist of an entire set of drawings which a client/owner can view with a free downloaf (similar to Adobe PDF files). (T.F)

10. It is not possible to remove sheets from the Drawing List without deleting that Sheet from the project. (T/F)

Self-Exam Answers:
1 - F, **2** - F, **3** - T, **4** - F, **5** - Sheet

Notes:

Notes:

Notes:

Notes:

Notes:

Notes:

Notes:

Notes:

Notes:

Notes:

Notes:

Notes:

Notes:

Index

A

Accurender Settings	13-6
Activate View	14-5
Align	5-11, 5-26, 10-7
Annotation	2-11, 8-35
Appears in Drawing List (sheet index)	14-12
Arc, Detail Lines	4-18, 4-25
Array	4-27, 5-53
Attach	7-15

B

Bath Room	11-1
Browser Organization	14-8
Bulkhead	8-23

C

Cabinets	11-2, 11-7
Callout	10-14
Camera	13-1, 13-20
Camera; Eye Elevation	13-23
Ceiling height	8-16
Ceiling Properties	8-22
Ceiling tool	8-16
Ceiling; Bulkhead	8-23
Ceiling; Deleting	8-20
Chain (option)	5-9
Circle (detail line)	3-15
Close Revit project	1-12
Component	8-30, 9-13, 11-20, 13-27
Component, Load	10-24
Control Arrows	7-7
Copy	3-26
Copy to Clipboard	4-23, 6-6, 6-19, 8-11
Counter Top	11-3, 11-10
Crop Region	9-5, 11-19, 13-3, 13-21, 13-26
Crossing Window	3-24

D

Daylights (rendering)	13-23
Deactivate View	14-6
Default 3D View	1-19, 5-21

Delete	3-22
Design bar	1-5, 12-4
Design Options	9-24, 14-9
Detail Components	10-20
Detail Level (view)	5-19, 7-1, 10-3
Detail Line	3-4, 3-11, 3-13, 3-20, 4-12
Dimension tool	2-14, 3-6, 6-34, 6-38
Dimensions; Activate	7-33
Dimensions; Locking	3-8, 5-13
Dimensions; Masonry	5-24, 5-36
Dimensions; Suffix	10-17
Dimensions; Witness Line	5-27
Dimension Snaps	3-17
Display Model	13-25
Door Tag	5-41, 12-4
Door tool	2-5, 5-35, 6-9
Door; Load from Library	5-37
Door; Properties	5-42
Door Schedule	12-6
Drafting View	3-3, 4-2, 4-4
3D DWF	14-15
Dynamically Modify View	1-18

E

Element Properties	5-48
Elevation	1-7, 6-4, 9-1, 9-18, 11-4, 11-13
Ellipse	4-21
Export Image	13-11, 13-25

F

Far Clip Plane	9-5, 10-2
Fillet Arc	4-14
Filter	5-19, 6-18
Fireplace	5-54
Flip Symbol (element control)	5-20, 5-42
Floor tool	8-1
Furniture	11-20
Footing, Continuous	10-12

H

Hide / Isolate	6-12, 11-22
Hide Annotation in View	9-19
Hide Views on Sheets	14-8

J

Join Geometry	6-8
Join/Unjoin Roof	7-18, 7-29

K

Kitchen Layout	11-7

L

Length (entering)	4-3, 9-15
Level	6-32, 7-21
Light Fixture	8-30, 8-34
Linework	11-9, 11-16
Load Family	5-36

M

Materials	8-14, 8-21
Menu bar	1-4
Mirror	4-26
Modify	3-5
Move	3-28, 4-5, 4-9, 4-23, 5-21

N

New (project)	1-13, 5-2

O

Offset	4-15
Open	1-8
Openings	5-45, 10-8
Options bar	1-5

P

Paste Aligned	6-6, 6-18
Print	2-15, 14-17
Project Browser	1-5, 5-4, 5-38
Project Browser Organization	5-52, 14-8
Project Information	5-5
Properties button	1-6

R

Radiosity (Radiate)	13-23
Railing (guardrail)	11-25, 11-29
Raster Images	14-14
Rectangle (dtl. line)	3-13
Redo	3-5
Region Raytrace	13-16
Rendering (raytrace)	13-10
Rendering material	13-4, 13-17
Rendering; RPC People	13-27
Repeating Detail	10-25
Resize	3-32, 4-22
Roof tool	2-9, 7-1
Roof; Flat	7-9
Roof; Gable	7-4, 7-11, 7-23
Roof; Hip	7-3, 7-28
Roof; Join/Unjoin	7-18, 7-29
Roof; Shed	7-6
Room Schedule	12-10
Room Tag	12-1
Room Tag; Properties	12-13
Rotate	3-28, 8-20

S

Save	1-12, 1-14, 3-15
Schedule/Quantities	12-6, 12-10
Scroll Wheel	1-19
Section	10-1
Section Box	13-13, 13-19
Section; Far Clip Plane	9-4
Selection Window	3-23
Shadows	13-15
Sheet	5-6, 14-1
Sheet Index	14-10
Skylight	7-32
Snap	2-8, 3-16
Solid Form	9-35
Split (wall)	5-29
Stair	6-20
Stair Width	6-24
Status bar	1-6, 5-11
Sun Settings	13-8, 13-16
Surface Pattern	8-21
Sweeps	5-32

T

TAB key	2-8, 8-11
Tag all not tagged	12-4
Tape Measure	2-4
Template	1-14, 5-2
Text Properties	8-37
Text Tool	2-11, 3-35, 8-35, 8-36, 10-18
Titleblock	14-1
Toolbars	1-4
Training Files	1-9
Trim	5-29
Type Parameters	5-42
Type Selector	1-6, 5-3

U

Underlay	6-1, 6-13
Undo	3-5
User Interface	1-3

V

View Control Bar	1-6
View Range	5-51
View Scale	2-12, 6-8, 9-7, 9-20, 10-15
Viewport	14-3
Visibility Graphics Overrides	8-5, 11-22

W

Wall; Footing	10-10
Wall Height	5-22, 6-3
Wall tool	2-2, 5-4, 5-8, 5-14, 5-23, 6-13
Wall tool; Base Offset	8-24
Wall tool; Chain	5-9
Wall tool; Edit Assembly	5-16, 5-31
Wed Library	5-38, 5-54
Window; Duplicate	5-50
Window menu	1-10
Window tool	5-46, 6-11
Window; Properties	5-49
Window; Sill Height	5-48

Z

Zoom	1-15

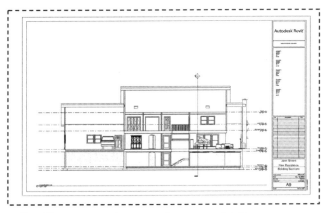

Sections *Sample image from page 14-7*

Renderings *Sample image from page 13-29*

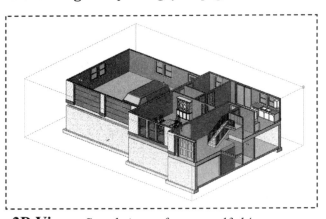

3D Views *Sample image from page 13-14*

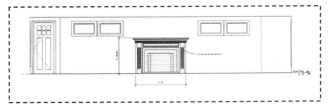

Interior Elevations *Sample image from page 9-22*